Thin-Films for Machining Difficult-to-Cut Materials

This book presents a balanced blend of fundamental research such as principles and characteristics of machining of difficult-to-cut materials and coating techniques and in-depth practical information on coatings techniques and classifications, the effect of coating parameters on machining responses, and finite element analysis of the machining performance of coated tools. In addition to the benefits of the thin-film deposition on the cutting tools, the limitations of the coating deposition techniques and the coating properties are also discussed.

Features:

- Associates the application of coating technology for improving machining characteristics of difficult-to-cut materials.
- Elaborates effect of coating architecture on the output machining parameters.
- Explores the performance of coated cutting tools.
- Discusses advanced coating systems and their application.
- Includes industrial case studies and practical implementations where coatings were applied for the machining of difficult-to-cut materials.

This book is aimed at researchers and graduate students in thin-films, coatings, machining, materials engineering, and manufacturing.

Advanced Materials Processing and Manufacturing

Series Editor: Kapil Gupta

The CRC Press Series in *Advanced Materials Processing and Man*ufacturing covers the complete spectrum of materials and manufacturing technology, including fundamental principles, theoretical background, and advancements. Considering the accelerated importance of advances for producing quality products for a wide range of applications, the titles in this series reflect the state-of-the-art in understanding and engineering the materials processing and manufacturing operations. Technological advancements for enhancement of product quality, process productivity, and sustainability, are on special focus including processing for all materials and novel processes. This series aims to foster knowledge enrichment on conventional and modern machining processes. Micro-manufacturing technologies such as micro-machining, micro-forming, and micro-joining, and Hybrid manufacturing, additive manufacturing, near net shape manufacturing, and ultra-precision finishing techniques are also covered.

Advanced Materials Characterization
Basic Principles, Novel Applications, and Future Directions
Ch Sateesh Kumar, M. Muralidhar Singh and Ram Krishna

Thin-Films for Machining Difficult-to-Cut Materials
Challenges, Applications, and Future Prospects
Ch Sateesh Kumar and Filipe Daniel Fernandes

For more information about this series, please visit: www.routledge.com/Advanced-Materials-Processing-and-Manufacturing/book-series/CRCAMPM

Thin-Films for Machining Difficult-to-Cut Materials

Challenges, Applications, and Future Prospects

Ch Sateesh Kumar and
Filipe Daniel Fernandes

CRC Press is an imprint of the
Taylor & Francis Group, an **informa** business

Designed cover image: Ch Sateesh Kumar

First edition published 2023
by CRC Press
6000 Broken Sound Parkway NW, Suite 300, Boca Raton, FL 33487-2742

and by CRC Press
4 Park Square, Milton Park, Abingdon, Oxon, OX14 4RN

CRC Press is an imprint of Taylor & Francis Group, LLC

© 2023 Ch Sateesh Kumar and Filipe Daniel Fernandes

Reasonable efforts have been made to publish reliable data and information, but the author and publisher cannot assume responsibility for the validity of all materials or the consequences of their use. The authors and publishers have attempted to trace the copyright holders of all material reproduced in this publication and apologize to copyright holders if permission to publish in this form has not been obtained. If any copyright material has not been acknowledged please write and let us know so we may rectify in any future reprint.

Except as permitted under U.S. Copyright Law, no part of this book may be reprinted, reproduced, transmitted, or utilized in any form by any electronic, mechanical, or other means, now known or hereafter invented, including photocopying, microfilming, and recording, or in any information storage or retrieval system, without written permission from the publishers.

For permission to photocopy or use material electronically from this work, access www.copyright.com or contact the Copyright Clearance Center, Inc. (CCC), 222 Rosewood Drive, Danvers, MA 01923, 978-750-8400. For works that are not available on CCC please contact mpkbookspermissions@tandf.co.uk

Trademark notice: Product or corporate names may be trademarks or registered trademarks and are used only for identification and explanation without intent to infringe.

ISBN: 9781032375120 (hbk)
ISBN: 9781032375137 (pbk)
ISBN: 9781003340553 (ebk)

DOI: 10.1201/9781003340553

Typeset in Times
by Deanta Global Publishing Services, Chennai, India

Contents

About the authors vii
Preface ix
Acknowledgments xi
Introduction xiii

Chapter 1 Introduction to machining 1
    1.1   Mechanism of chip formation 1
    1.2   Different metal-cutting operations 4
    1.3   Difficult-to-cut materials 7
    1.4   Cutting tool materials 10
    References 12

Chapter 2 Effect of coatings on machining parameters 17
    2.1   Machining difficulties and challenges 17
    2.2   Available remedies for machining difficulties 21
    2.3   Need for thin-films in machining 23
    References 23

Chapter 3 Coating technologies 27
    3.1   Physical vapor deposition 27
    3.2   Chemical vapor deposition 33
    3.3   Hybrid deposition techniques 39
    References 40

Chapter 4 Classification of coatings 41
    4.1   Generation of coatings 41
    4.2   Classification based on the coating material 43
    4.3   Classification based on coating properties 44
    4.4   Classification based on coating architecture/design 45
    References 47

Chapter 5 Application of coatings for machining 49
    5.1   Hard coatings 49
    5.2   Low-friction coatings 53

	5.3	Hybrid coatings..55
	5.4	Duplex coatings 55
	5.5	Other combinations 57
	5.6	Limitations of thin-films for machining of difficult-to-cut-materials..58
	References 60

Chapter 6 Effect of coatings on machining parameters......................................63

	6.1	Effect of coating thickness 63
	6.2	Effect of coating architecture/structure..................................69
	6.3	Performance of coated tools with lubrication.........................75
	References 79

Chapter 7 Significance of finite element analysis in analyzing performance of coated cutting tools 83

	7.1	Study of machining forces and stress distribution 84
	7.2	Study of temperature distribution...91
	7.3	Chip morphology analysis..93
	References 95

Chapter 8 Conclusions 99

Index..103

About the authors

Dr. Ch Sateesh Kumar completed his PhD at the National Institute of Technology, Rourkela, India. He worked as Visiting Assistant Professor at the Thapar Institute of Engineering and Technology, India. Later he joined the Advanced Materials Group at the Czech Technical University, Prague, Czech Republic as Postdoctoral Researcher. Presently he is working as a Senior Researcher at the Aeronautics Advanced Manufacturing Centre, Bilbao, Spain in the Department of Mechanical Engineering, Madanapalle Institute of Technology and Science, India. Also, he is working as a Senior Research Associate in the Department of Mechanical and Industrial Engineering Technology, University of Johannesburg, South Africa. His research interests include machining, surface modification techniques, coating deposition and characterization, and tribology. He has published 21 research articles in highly reputed journals and also presented his work at seven renowned conferences. As per Google Scholar, his h-index and i10-index are 10. He has also extended his services as a reviewer in many reputed journals.

Filipe Daniel Fernandes is currently an Assistant Professor at the School of Engineering at the Polytechnic of Porto, Portugal. In parallel, he integrated the body researcher group of IPN – Instituto Pedro Nunes, where he collaborated in the planning, preparation, and submission of project proposals to different funding entities. He is also a collaborator at the University of Coimbra, Portugal. In 2017 he was awarded a postdoc grant from FCT (SFRH/BPD/116334/2016) having as host institution the University of Minho, Portugal, where he improved the machinability of Ti alloys through the development of novel physical vapor deposition (PVD) films. He was placed in first place among 68 candidates. His work was recognized at the International Conference on Plasma Surface Engineering (Early Career Award 2018), perhaps the most known international conference on surface engineering held in Europe. In 2019 he won the prestigious Marie Curie fellowship and joined the Advanced Materials group at the Czech Technical University in Prague. Since 2010, Filipe Fernandes published 52 articles in international journals, being the first author in 17 of them (one as sole author) and in 21 manuscripts as the last author (four in in-press condition), showing a strong level leadership of his small group (currently one postdoc and one PhD; and one PhD already defended). According to Scopus, currently his h-index is 15 with a total of 607 citations. He is/was a member of the advisor committee/scientific commission of conferences, member of evaluation committees of research projects, member of scientific societies, member of the editorial board of scientific journals, and member organizer of seminars for high school students and graduate researchers. He participated in several national and European research projects; he is currently the PI of the MCTool21 project at the University of Coimbra, Portugal. He supervised and is supervising PhD and master's students. He was a jury member Jury for PhD and master's theses. This year he became a father.

Preface

Thin-films or surface coatings have been an integral part of metal cutting tools for improving their durability. The need for new materials with superior strength and wear resistance has also enhanced the demand for cutting tools with excellent wear resistance, and thermal and chemical stability. However, over the years it has been observed that the improvement in the physical properties of cutting tools with advanced cutting tool materials is not enough to meet the demand for high durability and superior performance. One of the main reasons behind this shortcoming is the cost associated with high-end tool materials like cubic boron nitrite (CBN), polycrystalline cubic boron nitride (PCBN), and diamond. However, the economic aspect of the cutting tools has been met to some extent by the use of ceramic and mixed ceramic cutting tools but they have their limitations due to their brittle nature.

On the contrary, it has been revealed by various researchers that the use of thin-film deposition on the cutting tools can significantly improve the durability and machining performance of cutting tools which is a much better economical solution than spending on cutting tools made of high-end materials. Thus, the present book introduces various machining processes and discusses their limitations, possible solutions, and the need for thin-films in machining. Furthermore, various coating techniques used for thin-film depositions on the cutting tools like chemical vapor deposition (CVD), and physical vapor deposition (PVD), and their types will be discussed. Also, the coating architecture/structure and its significance in relation to the machining performance will be discussed.

Later, the applicability of the coatings for machining different types of materials like hardened steels, superalloys, and other materials that come under the classification of difficult-to-cut materials will be examined. The main focus during the applicability study will be on the effect of exhibited properties and architecture of the deposited thin-films. Also, the book will elaborate in brief on the effect of coating thickness, chemical composition, and coating architecture on the machining output parameters. Finite element modeling and analysis is an important tool for predicting the machining performance of cutting tools and thus, at the end, the performance analysis of coated cutting tools and machining process using finite element analysis and modeling techniques are explained. In addition to the benefits of thin-film deposition on the cutting tools, the limitations of the coating deposition techniques and the coating properties will also be looked at.

Acknowledgments

To conclude, we would like to acknowledge the significant help that we have received from our colleagues, students, and family during the preparation of this book. We would like to acknowledge the sincere support of Prof. Kapil Gupta and Prof. Gagandeep Singh throughout the course of writing the book. We would also like to thank our departments; CFAA, Bilbao, Spain; Department of Mechanical Engineering, ISEP-Polytechnic of Porto, Portugal; and Department of Mechanical and Industrial Engineering Technology, University of Johannesburg, South Africa for their motivation and support.

Introduction

This book aims at bridging the gap between thin-film characterizations and the application of thin-films for machining difficult-to-cut materials. The book initially introduces the metal cutting operations, chip formation mechanism, cutting tool materials, and different difficult-to-cut materials so as to familiarize the reader with the terminology regarding the metal cutting operation in the subsequent sections. Furthermore, the book speaks about various machining challenges and difficulties taking into account the machining of materials like nickel-based superalloys and hardened steels which are considered difficult to cut.

Later, the coating deposition techniques, namely chemical vapor deposition (CVD), physical vapor deposition (PVD), and hybrid deposition processes are elaborated on. This introduction to coating deposition techniques is necessary to understand the properties and behavior of thin-films and their performance during the metal cutting operation. Also, a detailed classification of thin-films has been presented based on the generation, chemical composition, coating architecture/structure, and properties. After the introduction to coating deposition techniques and different coatings, the application of coatings during the machining of difficult-to-cut materials is discussed with the help of examples. The application of thin-films is explored with regard to the applicability of hard and low friction, self-lubricating, hybrid, duplex, and other advanced coating systems during metal cutting processes.

The book gives insight into specific cases of thin-films such as the effect of coating thickness and coating architecture/structure on the performance of coatings which has been elaborated by giving proper illustrations. Further, hybrid systems that use coolants to provide lubrication and cleaning during machining with coated tools have been discussed. Also, the significance of finite element analysis in investigating the performance of coated cutting tools has been elaborated. The book provides comprehensive knowledge about thin-film application with the help of high-quality peer-reviewed published work and thus, it would be useful for UG and PG students, PhD scholars, and researchers interested in the field of thin-films and their application.

1 Introduction to machining

The machining process is one of the most significant parts of the manufacturing process industry when it comes to giving shape to the stock by removing excess material using a metal-cutting operation. When it comes to machining which involves high temperatures and machining forces, the cutting tool's durability and the quality of the machined surface become the most significant parameters to consider after completion of the process (Chinchanikar & Choudhury, 2014; Outeiro et al., 2008). Also, irrespective of the type of machining process, the outcomes of any metal-cutting operation are highly dependent on the type of material to be machined, the cutting tool material, and obviously the machining conditions and parameters. However, if we just consider the effect of tool and workpiece material, an optimized combination of the workpiece and cutting tools is extremely necessary (El Hakim et al., 2011; Kamata & Obikawa, 2007; Kumar & Patel, 2018a). Thus, in the subsequent sections, the metal-cutting process mechanism, different hard-to-cut workpiece materials, and different cutting tool materials will be discussed in detail.

1.1 MECHANISM OF CHIP FORMATION

As already discussed, the machining process involves removal of excess material from the workpiece materials that are termed "chips". These chips are the waste material in the machining process. However, the mechanism of the formation of chips which depends on the properties of the workpiece material, machining conditions, and process parameters is quite significant during the machining operation (An et al., 2014; Kim et al., 2016; K. Li et al., 2002). For the chip formation mechanism, the cutting tool exerts a compressive force onto the workpiece as shown in Figure 1.1 and when this compressive force generates more stress than the yield strength of the material, the latter's removal takes place due to shearing. The applied compressive force that causes shearing is known as the "cutting force". This force creates a plane of deformation due to the localized plastic deformation called the "shear plane" along which the material removal takes place (Globo et al., 1975). Again, the type of chips formed during any machining operation is highly dependent on the properties of the workpiece material (hard, brittle, ductile, etc.), cutting conditions (dry machining or machining with coolant), and machining parameters (depth of cut, feed rate, cutting speed) (Baeker & Martin, 2015; Binder et al., 2015). The chips formed during any machining process can mainly be of three types: either continuous,

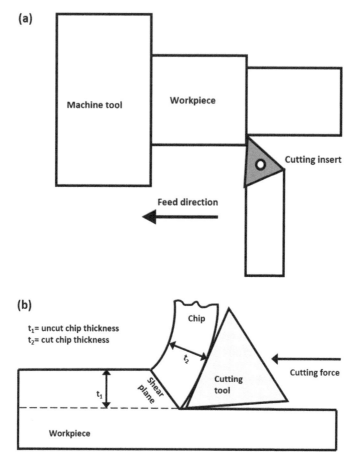

FIGURE 1.1 Schematic representation of (a) turning process and (b) chip formation mechanism

discontinuous, or continuous, with built-up edges (see Figure 1.2). Continuous chips are in the form of long curled structures and are basically generated while machining ductile materials with sufficient plastic deformation capability. Also, the formation of these chips is facilitated by high cutting speeds and low feed rates. Further, discontinuous chips are formed while machining materials that are very brittle undergo deformation, and thus, the formed chips fracture to form small segments. Next, when the cutting temperatures and forces are very high, the subsequent pressure at the cutting edge causes the welding of the workpiece material to the cutting tool. These loosely welded particles can be carried away by the moving chips. This phenomenon of a built-up-edge (BUE) formation can be attributed to high cutting forces, temperature, and the chemical affinity of the tool and the workpiece material (Globo et al., 1975; Kümmel et al., 2015; Sui et al., 2015; Ucun et al., 2015; Zhvirblis, 1987).

Introduction to machining

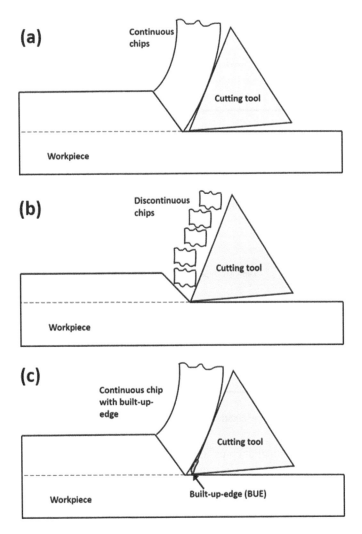

FIGURE 1.2 Types of chips formed while machining (a) continuous, (b) discontinuous, and (c) continuous chips with built-up-edge (BUE)

Further, there can be some modifications of the chip shapes and forms based on the materials that are being machined and the machining parameters. Figure 1.3 shows the formation of chips with serrated teeth or saw teeth or segmentations (Jomaa et al., 2017). These chips are formed when the deformation in the shear zone exceeds the limiting strength causing high shear instability (Globo et al., 1975; Jomaa et al., 2017). Also, it has been observed that the serrated chips are formed during high-speed machining, machining of difficult-to-cut materials like superalloys and hardened steels, and also while using high negative rake angles (Jomaa et al., 2017; Ohbuchi & Obikawa, 2003; Wang et al., 2013). In addition, there is a special

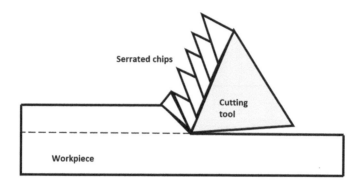

FIGURE 1.3 Schematic representation of chip formation with serrated teeth

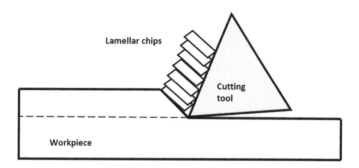

FIGURE 1.4 Schematic representation of lamellar chip formation

class of chips called "lamellar chips" (see Figure 1.4) that are basically formed when the deformation in the shear zone causes a reduction in the strength of the material (Globo et al., 1975). These chips are mostly like chips with serrated teeth but are formed mostly while machining low thermal conductivity non-crystalline materials at lower cutting speeds due to thermal instability at the chip deformation zone (Jiang & Dai, 2009).

1.2 DIFFERENT METAL-CUTTING OPERATIONS

The removal of material from the blank to give it a desired shape can be performed using various techniques and processes. The selection of the metal-cutting process depends upon the complexity of geometry, surface quality, and dimensional accuracy of the final product. Figure 1.5 shows the various metal-cutting operations with examples that are used in the manufacturing industry for the production of various mechanical components (Globo et al., 1975; Zhvirblis, 1987). However, thin-films can only be employed in metal-cutting operations that involve material removal with the help of cutting tools that come in direct contact with the workpiece (Bouzakis et al., 2012; Endrino et al., 2006; Kurniawan et al., 2010; Sateesh Kumar & Kumar Patel, 2017). Thus, the present section will discuss various

Introduction to machining

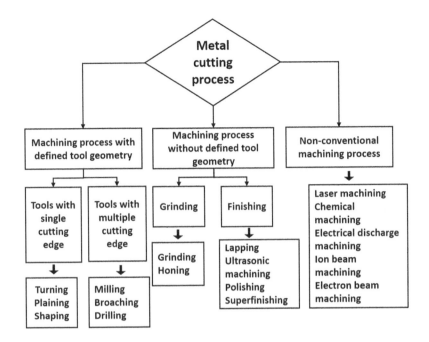

FIGURE 1.5 Classification of metal-cutting processes

machining techniques in which the cutting tool comes into direct contact with the workpiece thereby plastically deforming and removing the material by shearing. These metal-cutting operations can be classified based on the movement of the cutting tool and the workpiece. Again, based on the requirement for the final product's shape, dimensional accuracy, and surface quality, the motion of the cutting tool can either be rotatory (e.g., drilling and milling) or translatory (e.g., turning and shaping). In the same way, the workpiece that has to be given desired shape by material removal can translate, rotate, or remain stationary depending on the requirement and the metal-cutting operation used. However, every machine tool depicting a certain metal-cutting operation may have certain movement restrictions for both tool and the workpiece. However, the metal-cutting machine tools have been continuously modified to get more degree of freedom so that highly complex structures can be developed using machining (e.g., 5-axis CNC lathe or milling machine). A few examples of these metal-cutting operations are turning, milling, drilling, shaping, broaching, and plaining. These cutting tools basically have defined tool geometry. However, the cutting process can be carried out using a single cutting edge (e.g., turning) or using multiple cutting edges (e.g., milling). In both cases, the machining performance can be enhanced by depositing thin-films on the cutting tools. The deposition of coatings can act as a thermal barrier between the tool substrate and the workpiece, lubricant by forming tribo-lubricious phases, and can also improve the wear resistance of the cutting tools during machining operation (Chen et al., 2010; El Hakim et al., 2011; Kumar & Patel, 2018a; Kumar Sahoo & Sahoo, 2013; Martinez et al., 2017; Sateesh et al., 2020).

Another important phenomenon apart from the complexity of the final product is the machinability of the workpiece which is highly influenced by the chemical, mechanical and thermal properties of the workpiece material (Chinchanikar & Choudhury, 2015; Molaiekiya et al., 2020). The metal-cutting operations with defined cutting edges have a similar sequence of operations and final investigation stages irrespective of the machine tool, cutting tool, and machine tool used in the process. Figure 1.6 shows a work flow-chart of initiation (selecting cutting tool, workpiece material, and cutting environment either dry or with coolant) and final investigations (studying tool life, tool wear, machining forces, and surface roughness) that are basically carried out in a metal-cutting operation using cutting tools with defined cutting edges (Globo et al., 1975). As the machinability is influenced by the properties of the workpiece material, the cutting tools, and the cutting environment has to be selected judiciously. Further, the tool wear, tool life, surface roughness of the machined surface, chip morphology, and machining forces also get significantly influenced by the workpiece and cutting tool material properties. Thus, it becomes very necessary to understand various materials used for cutting tools and different workpiece materials that are difficult to cut.

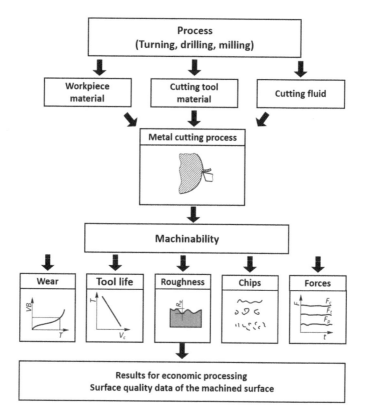

FIGURE 1.6 Work flow-chart of a machining process from initial material selection to the final product development

Introduction to machining

1.3 DIFFICULT-TO-CUT MATERIALS

The chemical, mechanical and thermal properties of different materials make them difficult to cut under different situations. The demand for advanced materials in challenging sectors like defense, aerospace, battery manufacturing, medical equipment, and the manufacturing sector dealing with high-temperature applications with superior wear resistance, heat resistance, and chemical and thermal stability under adverse conditions led to the development of materials like hardened steels, titanium alloys, superalloys, new ceramics and composites (An et al., 2021; Daymi et al., 2009; Kumar & Patel, 2017; Tandon et al., 1990; Thakur & Gangopadhyay, 2016). The various difficult-to-cut materials have been classified and are listed in Figure 1.7. The different materials will have varying levels of machinability depending on their properties. For instance, when the material hardness increases, the machining forces for material removal also increase resulting in a subsequent increase in cutting temperatures. These adverse conditions will deteriorate the tool more aggressively both mechanically (e.g., by abrasion) and thermally (e.g., diffusion or chemical reaction at high temperature like oxidation) when compared to the machining of a reasonably softer material (Kumar & Patel, 2018b). Thus, each material class will be discussed one by one to provide more clarity.

1.3.1 HARDENED STEELS

The need to use tougher and harder steels for various applications is increasing rapidly due to advances in manufacturing systems and the demand for improved parts with a higher wear resistance for prolonged durability (Chinchanikar & Choudhury,

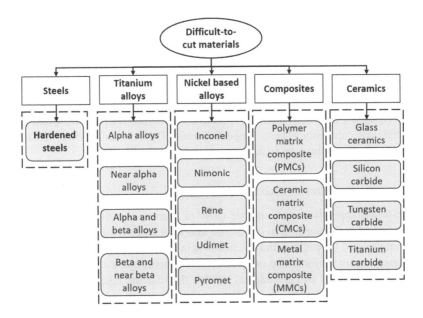

FIGURE 1.7 Classification of different difficult-to-cut materials

2015). Hardened steels are basically quenched to increase their hardness and steels with hardness above 45 HRC are generally considered hard steels. However, this situation has changed over the years with advances in cutting tool materials for machining applications (Aslantas et al., 2012). The hardness of the steels increases after quenching which further enhances the need for higher machining forces for the removal of material thereby increasing the cutting temperatures for the same machining conditions. Thus, higher hardness will proliferate tool wear through abrasion and thermal deterioration (Sateesh Kumar & Kumar Patel, 2017).

1.3.2 Titanium alloys

The aerospace industry stresses the use of materials with low density and high strength to keep the loads at elevations at a minimum. This becomes more imperative due to increasing fuel costs. Titanium alloys have proved to be a favorite choice in this regard due to their higher specific weight-to-strength ratio and resistance to corrosion when compared to components made of steel or aluminum (Hosseini & Kishawy, 2014). This phenomenon becomes more significant at higher temperatures. When it comes to classification, the titanium alloys can be classified into four groups. These four groups are alpha alloys, near alpha alloys, alpha and beta alloys, and beta and near beta alloys. Titanium has two allotropes, namely, alpha phase and beta phase. The alpha phase is a low-temperature allotrope whereas the beta phase is a high-temperature body-centered cubic allotrope formed over 882°C. Further, the addition of alloying elements like aluminum (Al) and tin (Sn) can increase the transformation temperature. These elements are termed "alpha-stabilizers". On the other hand, when the addition of elements reduces transformation temperature, the elements are termed "beta-stabilizers". Molybdenum (Mo) and vanadium (V) are examples of such elements. When alpha-stabilizers are present with minor beta stabilizers as alloying elements, a near alpha titanium alloy is formed. However, when alpha and beta phases are present together, an alpha-beta titanium alloy is formed. In the same way there can be a near beta and beta phase with beta stabilizers (Boyer, 1996; Brewer et al., 1998; Ezugwu et al., 2003; Honnorat, 1996). The advantages offered by titanium alloys are negated by their properties like low thermal conductivity and elastic modulus, enhanced chemical reactivity at higher temperatures, and higher hardenability capability. These properties reduce the machinability of titanium alloys, categorizing them as difficult-to-cut materials.

1.3.3 Nickel-based superalloys

Nickel-based superalloys are another class of materials with a high strength-to-weight ratio when compared to aluminum and steel and thus, it fulfills almost 50% material demand for engines and gas turbines in the aerospace industry. Further, the high-temperature performance of nickel-based superalloys with superior resistance to corrosion, thermal and mechanical shocks, thermal and mechanical fatigue, erosion, and creep makes them an excellent choice for use in nuclear reactors, marine equipment, the petrochemical industry, and other high-temperature applications (Ezugwu et al., 2003). Some commercially available nickel-based superalloys are

TABLE 1.1
Examples of a few commercially available nickel-based super alloys

Inconel (587, 597, 600, 601, 617, 625, 706, 718, X750, 901)

Unitemp AF2-IDA6 Cabot 214 Haynes 230

  Nimonic (75, 80A, 90, 105, 115, 263, 942, PE 11, PE 16, PK 33, C-263)

  Rene (41, 95)

  Udimet (400, 500, 520, 630, 700, 710, 720)

  Pyromet 860

  Astroloy

  Waspaloy

Inconel, Nimonic, Pyromet, Udimet, and Rene (Choudhury & El-Baradie, 1998) which are also listed in Table 1.1. These alloys are essentially a combination of 38 to 76 wt.% of nickel (Ni), a maximum of 27 wt.% chromium (Cr), and a maximum of 20 wt.% cobalt (Co). However, their oxidation properties can be enhanced by the addition of molybdenum (Mo), tantalum (Ta), and tungsten (W) (Ezugwu et al., 2003). The high-temperature resistance properties of nickel-based superalloys make them extremely difficult to cut leading to high tool wear and low tool life.

1.3.4 Composites

A composite is a material that is formed by mixing two or more constituent elements to form a new material whose properties are entirely different from the initial elements. In a composite both the materials are present in their original form which differentiates them from mixtures and solid solutions. The composites can be classified basically into three categories namely; polymer matrix composites (PMCs), metal matrix composites (MMCs), and ceramic matric composites (CMCs). The term matrix refers to the main constituent material to which other constituents are added to form a composite. The added constituents are often termed as reinforcements (Wikipedia contributors, 2022b). Thus, if a constituent is added to a metal matrix, then the resultant composite is called a "metal matrix composite". The same statement can be made for polymer and ceramic matrix composites. These composites may exhibit high hardness and brittleness, high strength-to-weight ratio, and low thermal expansion coefficient (Garg et al., 2010) which makes them extremely difficult to machine.

1.3.5 Ceramics

Ceramics are materials that are formed at high temperatures by heating inorganic nonmetallic compounds (Wikipedia contributors, 2022a). These ceramics possess high hardness, melting point, thermal resistance, and corrosion resistance which makes them extremely suitable for various strategic applications (Y. Liu et al., 2017). However, ceramics are also extremely brittle, making their plastic deformation almost impossible. Further, due to high melting points and low thermal softening capabilities, ceramics are extremely difficult to machine. Some common examples of ceramics are tungsten carbide (WC), titanium carbide (TiC), silicon carbide (SiC), titanium aluminum nitride (TiAlN), and so on (Dobrzański & Mikuła, 2005; Eblagon et al., 2007; W. Liu et al., 2017). Thus, from the above examples, it can be concluded that even compounds of carbon can be considered ceramics. Also, due to the very high hardness of ceramics, they are rarely machined using conventional metal-cutting processes.

1.4 CUTTING TOOL MATERIALS

The material for cutting tools should be selected judiciously based on the applicability, precision, and surface quality requirements for the metal-cutting process. As a general rule, the material of the cutting tool should be harder than the material to be cut. Further, while machining difficult-to-cut materials, the prerequisites for the cutting tools are high hot hardness, chemical and thermal stability at high temperatures, and high toughness (Kumar & Patel, 2017; Sateesh Kumar & Kumar Patel, 2017). Figure 1.8 shows different types of cutting tool materials as a function of hardness and toughness. It is evident from the illustration that for cutting tools, toughness and hardness have an inverse relationship. Cutting tools with higher hardness like those with polycrystalline diamond (PCD) exhibit lower toughness. Further, the cutting tools made of high-speed steel are extremely tough but at the same time exhibit very low levels of hardness when compared to other harder cutting tools made of polycrystalline cubic boron nitrides (PCBN) and ceramics. Thus, it is essential to understand the use of different cutting tool materials for machining difficult-to-cut materials.

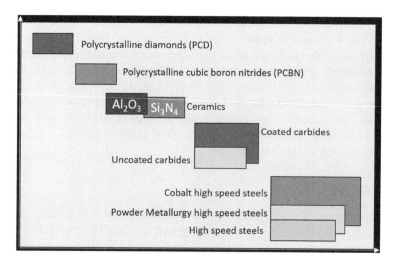

FIGURE 1.8 Classification of different cutting tool materials as a function of toughness and hardness

Introduction to machining

1.4.1 HIGH-SPEED STEELS (HSS)

High-speed steels (HSS) are cutting tool materials that possess high toughness which makes them extremely suitable for discontinuous or intermittent cutting applications like milling, drilling, broaching, and tapping (Hosseini & Kishawy, 2014). However, they are not the best choice when it comes to machining difficult-to-cut materials. Even though they are extremely tough, they cannot be used above 500 °C due to their low softening point at 600 °C (Ezugwu et al., 2003; Hosseini & Kishawy, 2014). On the contrary, the machining of difficult-to-cut materials leads to the generation of very high cutting temperatures (Kumar & Patel, 2017; Özel & Ulutan, 2012) which makes HSS cutting tools unsuitable for machining these materials.

1.4.2 CARBIDES

Carbides are another class of materials that are used for manufacturing cutting tools. Generally, two grades of carbides, namely, pure and mixed are used for commercial machining applications. Pure carbides consist of tungsten carbide (WC) with a small amount of cobalt (Co) ranging from 5 to 12 wt.%. On the contrary, the mixed grade consists of titanium carbide (TiC), tantalum carbide (TaC), or niobium carbide (NbC), in addition to tungsten carbide. These compounds are added to enhance the hardness, toughness, and other properties of pure tungsten carbide (Ezugwu et al., 2003; Hosseini & Kishawy, 2014). Cutting tools made of carbides with or without coating are widely used for machining difficult-to-cut materials (A.P. & V.G., 2015; L. Li et al., 2002).

1.4.3 CERAMICS

Ceramics are another important class of materials that are used for machining difficult-to-cut materials. They are more significant because of their lower chemical reactivity, higher hot hardness, and high thermal softening points (Koseki et al., 2017; Molaiekiya et al., 2020; Shalaby et al., 2014). A general classification of ceramic materials that are currently used for manufacturing cutting tools has been listed in Figure 1.9. Ceramic cutting tools materials are basically subdivided into alumina- (Al_2O_3) based oxide ceramics and silicon nitride- (Si_3N_4) based non-oxide

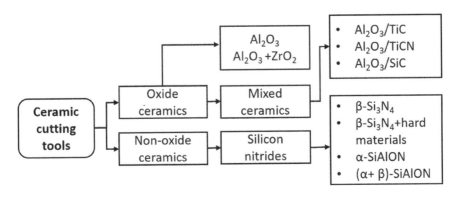

FIGURE 1.9 Classification of ceramic cutting tool materials

ceramics. Further, the Al_2O_3 ceramics have a subdivision of mixed ceramics which are mostly ceramic-ceramic composites like Al_2O_3/TiC, Al_2O_3/TiCN, and Al_2O_3/SiC. This modification helps in reducing the brittleness of Al_2O_3 oxide ceramics (Kumar & Patel, 2018a; Sateesh Kumar & Kumar Patel, 2017).

1.4.4 CUBIC BORON NITRIDES (CBN), POLYCRYSTALLINE CUBIC BORON NITRIDES (PCBN), AND POLYCRYSTALLINE DIAMONDS (PCD)

Cubic boron nitride (CBN) is a material that has hardness next to diamond. It also possesses better oxidation resistance when compared to diamond which starts graphitizing at 900°C whereas CBN remains chemically stable up to 2000°C with no signs of oxidation. Further, when the CBN grains are sintered at high temperature and pressure, a new class of CBN is formed, namely polycrystalline cubic boron nitride (PCBN) (Hosseini & Kishawy, 2014; Klocke & Kuchle, 2001). The CBN-based cutting tools exhibit high hot hardness and superior chemical stability at high temperatures. Another class of extremely hard cutting tool material is polycrystalline diamonds (PCD) that are formed by sintering diamond particles with a metallic binder at high temperature and pressure. PCDs exhibit high hot hardness and chemical stability but also have high reactivity with iron which makes them unsuitable for machining steels (Becker et al., 2015; Hosseini & Kishawy, 2014).

REFERENCES

A.P., K., & V.G., S. (2015). Characterization and performance of AlTiN, AlTiCrN, TiN/TiAlN PVD coated carbide tools while turning SS 304. *Materials and Manufacturing Processes, 30*(6), 748–755. https://doi.org/10.1080/10426914.2014.984217

An, Q., Chen, J., Ming, W., & Chen, M. (2021). Machining of SiC ceramic matrix composites: A review. *Chinese Journal of Aeronautics, 34*(4), 540–567. https://doi.org/10.1016/j.cja.2020.08.001

An, Q., Wang, C., Xu, J., Liu, P., & Chen, M. (2014). Experimental investigation on hard milling of high strength steel using PVD-AlTiN coated cemented carbide tool. *International Journal of Refractory Metals and Hard Materials, 43,* 94–101. https://doi.org/10.1016/j.ijrmhm.2013.11.007

Aslantas, K., Ucun, T. I., & Çicek, A. (2012). Tool life and wear mechanism of coated and uncoated Al 2O 3/TiCN mixed ceramic tools in turning hardened alloy steel. *Wear, 274–275,* 442–451. https://doi.org/10.1016/j.wear.2011.11.010

Baeker, M., & Martin, B. (2015). Finite element simulation of high-speed cutting forces. *Journal of Materials Processing Technology Finite Element Simulation of High-speed Cutting Forces. 176*(JUNE 2006), 117–126. https://doi.org/10.1016/j.jmatprotec.2006.02.019

Becker, F. G., Cleary, M., Team, R. M., Holtermann, H., The, D., Agenda, N., Science, P., Sk, S. K., Hinnebusch, R., Hinnebusch, A. R., Rabinovich, I., Olmert, Y., Uld, D. Q. G. L.. Q., Ri, W. K. H. U., Lq, V., Frxqwu, W. K. H., Zklfk, E., Edvhg, L. V., Wkh, R. Q., ... ﻃﺎﻟﺒﻲ, ف. (2015). No 主観的健康感を中心とした在宅高齢者における健康関連指標に関する共分散構造分析Title. *Syria Studies, 7*(1). https://www.researchgate.net/publication/269107473_What_is_governance/link/548173090cf22525dcb61443/download%0Ahttp://www.econ.upf.edu/~reynal/Civil   wars_12December2010.pdf%0Ahttps://think-asia.org/handle/11540/8282%0Ahttps://www.jstor.org/stable/41857625

Binder, M., Klocke, F., & Lung, D. (2015). Tool wear simulation of complex shaped coated cutting tools. *Wear, 330–331,* 600–607. https://doi.org/10.1016/j.wear.2015.01.015

Bouzakis, K. D., Pappa, M., Gerardis, S., & Skordaris, G. (2012). Determination of PVD coating mechanical properties by nanoindentations and impact tests at ambient and elevated temperatures. *Journal of the Balkan Tribological Association, 18*(2), 259–270.

Boyer, R. R. (1996). An overview of the use of titanium in the aerospace industry. *Materials Science and Engineering: A, 213*(1), 103–114. https://doi.org/10.1016/0921-5093(96)10233-1

Brewer, W. D., Bird, R. K., & Wallace, T. A. (1998). Titanium alloys and processing for high speed aircraft. *Materials Science and Engineering: A, 243*(1), 299–304. https://doi.org/10.1016/S0921-5093(97)00818-6

Chen, L., Wang, S. Q., Du, Y., Zhou, S. Z., Gang, T., Fen, J. C., Chang, K. K., Li, Y. W., & Xiong, X. (2010). Machining performance of Ti-Al-Si-N coated inserts. *Surface and Coatings Technology, 205*(2), 582–586. https://doi.org/10.1016/j.surfcoat.2010.07.043

Chinchanikar, S., & Choudhury, S. K. (2014). Hard turning using HiPIMS-coated carbide tools: Wear behavior under dry and minimum quantity lubrication (MQL). *Measurement: Journal of the International Measurement Confederation, 55,* 536–548. https://doi.org/10.1016/j.measurement.2014.06.002

Chinchanikar, S., & Choudhury, S. K. (2015). Machining of hardened steel – Experimental investigations, performance modeling and cooling techniques: A review. *International Journal of Machine Tools and Manufacture, 89,* 95–109. https://doi.org/10.1016/j.ijmachtools.2014.11.002

Choudhury, I. A., & El-Baradie, M. A. (1998). Machinability of nickel-base super alloys: A general review. *Journal of Materials Processing Technology, 77*(1), 278–284. https://doi.org/10.1016/S0924-0136(97)00429-9

Daymi, A., Boujelbene, M., Salem, S. Ben, Hadj Sassi, B., Torbaty, S., & Sassi, B. H. (2009). Effect of the cutting speed on the chip morphology and the cutting forces. *Manufacturing and Processing of Engineering Materials, 78. 1*(2), 77–83.

Dobrzański, L. A., & Mikuła, J. (2005). Structure and properties of PVD and CVD coated Al2O3 + TiC mixed oxide tool ceramics for dry on high speed cutting processes. *Journal of Materials Processing Technology, 164–165,* 822–831. https://doi.org/10.1016/j.jmatprotec.2005.02.089

Eblagon, F., Ehrle, B., Graule, T., & Kuebler, J. (2007). Development of silicon nitride/silicon carbide composites for wood-cutting tools. *Journal of the European Ceramic Society, 27*(1), 419–428. https://doi.org/10.1016/j.jeurceramsoc.2006.02.040

El Hakim, M. A., Abad, M. D., Abdelhameed, M. M., Shalaby, M. A., & Veldhuis, S. C. (2011). Wear behavior of some cutting tool materials in hard turning of HSS. *Tribology International, 44*(10), 1174–1181. https://doi.org/10.1016/j.triboint.2011.05.018

Endrino, J. L., Fox-Rabinovich, G. S., & Gey, C. (2006). Hard AlTiN, AlCrN PVD coatings for machining of austenitic stainless steel. *Surface and Coatings Technology, 200*(24), 6840–6845. https://doi.org/10.1016/j.surfcoat.2005.10.030

Ezugwu, E. O., Bonney, J., & Yamane, Y. (2003). An overview of the machinability of aeroengine alloys. *Journal of Materials Processing Technology, 134*(2), 233–253. https://doi.org/10.1016/S0924-0136(02)01042-7

Garg, R. K., Singh, K. K., Sachdeva, A., Sharma, V. S., Ojha, K., & Singh, S. (2010). Review of research work in sinking EDM and WEDM on metal matrix composite materials. *The International Journal of Advanced Manufacturing Technology, 50*(5–8), 611–624. https://doi.org/10.1007/s00170-010-2534-5

Globo, G., Kramar, D., & Kopa, J. (1975). *Metal Cutting.*-Theory and Applications. ISBN: 978-961-6536-85-1, UNIVERSITY OF BANJA LUKA FACULTY OF MECHANICAL ENGINEERING

Honnorat, Y. (1996). Issues and breakthrough in the manufacture of turboengine titanium parts. *Materials Science and Engineering: A, 213*(1), 115–123. https://doi.org/10.1016/0921-5093(96)10229-X

Hosseini, A., & Kishawy, H. A. (2014). *Machining of Titanium Alloys* (J. P. Davim (ed.)). Springer, Berlin Heidelberg. https://doi.org/10.1007/978-3-662-43902-9

Jiang, M. Q., & Dai, L. H. (2009). Formation mechanism of lamellar chips during machining of bulk metallic glass. *Acta Materialia, 57*(9), 2730–2738. https://doi.org/10.1016/j.actamat.2009.02.031

Jomaa, W., Mechri, O., Lévesque, J., Songmene, V., Bocher, P., & Gakwaya, A. (2017). Finite element simulation and analysis of serrated chip formation during high–speed machining of AA7075–T651 alloy. *Journal of Manufacturing Processes, 26*(October), 446–458. https://doi.org/10.1016/j.jmapro.2017.02.015

K??mmel, J., Braun, D., Gibmeier, J., Schneider, J., Greiner, C., Schulze, V., & Wanner, A. (2015). Study on micro texturing of uncoated cemented carbide cutting tools for wear improvement and built-up edge stabilisation. *Journal of Materials Processing Technology, 215*, 62–70. https://doi.org/10.1016/j.jmatprotec.2014.07.032

Kamata, Y., & Obikawa, T. (2007). High speed MQL finish-turning of Inconel 718 with different coated tools. *Journal of Materials Processing Technology, 192–193*, 281–286. https://doi.org/10.1016/j.jmatprotec.2007.04.052

Kim, D. M., Lee, I., Kim, S. K., Kim, B. H., & Park, H. W. (2016). Influence of a micropatterned insert on characteristics of the tool-workpiece interface in a hard turning process. *Journal of Materials Processing Technology, 229*, 160–171. https://doi.org/10.1016/j.jmatprotec.2015.09.018

Klocke, F., & Kuchle, A. (2001). *Manufacturing Process 1*. Publisher: Springer. ISSN 1865-0899 e-ISSN 1865-0902 ISBN 978-3-642-11978-1 e-ISBN 978-3-642-11979-8 DOI 10.1007/978-3-642-11979-8

Koseki, S., Inoue, K., Sekiya, K., Morito, S., Ohba, T., & Usuki, H. (2017). Wear mechanisms of PVD-coated cutting tools during continuous turning of Ti-6Al-4V alloy. *Precision Engineering, 47*, 434–444. https://doi.org/10.1016/j.precisioneng.2016.09.018

Kumar, C. S., & Patel, S. K. (2017). Surface & Coatings Technology Experimental and numerical investigations on the effect of varying AlTiN coating thickness on hard machining performance of Al 2 O 3 -TiCN mixed ceramic inserts. *SCT, 309*, 266–281. https://doi.org/10.1016/j.surfcoat.2016.11.080

Kumar, C. S., & Patel, S. K. (2018a). Investigations on the effect of thickness and structure of AlCr and AlTi based nitride coatings during hard machining process. *Journal of Manufacturing Processes, 31*, 336–347. https://doi.org/10.1016/j.jmapro.2017.11.031

Kumar, C. S., & Patel, S. K. (2018b). Performance analysis and comparative assessment of nano-composite TiAlSiN/TiSiN/TiAlN coating in hard turning of AISI 52100 steel. *Surface and Coatings Technology, 335*(September 2017), 265–279. https://doi.org/10.1016/j.surfcoat.2017.12.048

Kumar Sahoo, A., & Sahoo, B. (2013). Performance studies of multilayer hard surface coatings (TiN/TiCN/Al2O3/TiN) of indexable carbide inserts in hard machining: Part-II (RSM, grey relational and techno economical approach). *Measurement, 46*(8), 2868–2884. https://doi.org/10.1016/j.measurement.2012.09.023

Kurniawan, D., Yusof, N. M., & Sharif, S. (2010). Hard machining of stainless steel using wiper coated carbide: Tool life and surface integrity. *Materials and Manufacturing Processes, 25*(6), 370–377. https://doi.org/10.1080/10426910903179930

Li, K., Gao, X. L., & Sutherland, J. W. (2002). Finite element simulation of the orthogonal metal cutting process for qualitative understanding of the effects of crater wear on the chip formation process. *Journal of Materials Processing Technology, 127*(3), 309–324. https://doi.org/10.1016/S0924-0136(02)00281-9

Li, L., He, N., Wang, M., & Wang, Z. G. (2002). High speed cutting of Inconel 718 with coated carbide and ceramic inserts. *Journal of Materials Processing Technology*, *129*(1–3), 127–130. https://doi.org/10.1016/S0924-0136(02)00590-3

Liu, W., Chu, Q., Zeng, J., He, R., Wu, H., Wu, Z., & Wu, S. (2017). PVD-CrAlN and TiAlN coated Si3N4 ceramic cutting tools -1. Microstructure, turning performance and wear mechanism. *Ceramics International*, April, 0–1. https://doi.org/10.1016/j.ceramint.2017.04.041

Liu, Y., Deng, J., Wu, F., Duan, R., Zhang, X., & Hou, Y. (2017). Wear resistance of carbide tools with textured flank-face in dry cutting of green alumina ceramics. *Wear*, *372–373*, 91–103. https://doi.org/10.1016/j.wear.2016.12.001

Martinez, I., Tanaka, R., Yamane, Y., Sekiya, K., Yamada, K., Ishihara, T., & Furuya, S. (2017). Wear mechanism of coated tools in the turning of ductile cast iron having wide range of tensile strength. *Precision Engineering*, *47*, 46–53. https://doi.org/10.1016/j.precisioneng.2016.07.003

Molaiekiya, F., Aramesh, M., & Veldhuis, S. C. (2020). Chip formation and tribological behavior in high-speed milling of IN718 with ceramic tools. *Wear*, *446–447*. https://doi.org/10.1016/j.wear.2020.203191

Ohbuchi, Y., & Obikawa, T. (2003). Finite element modeling of chip formation in the domain of negative rake angle cutting. *Journal of Engineering Materials and Technology*, *125*(July 2003), 324. https://doi.org/10.1115/1.1590999

Outeiro, J. C., Pina, J. C., M'Saoubi, R., Pusavec, F., & Jawahir, I. S. (2008). Analysis of residual stresses induced by dry turning of difficult-to-machine materials. *CIRP Annals – Manufacturing Technology*, *57*(1), 77–80. https://doi.org/10.1016/j.cirp.2008.03.076

Özel, T., & Ulutan, D. (2012). Prediction of machining induced residual stresses in turning of titanium and nickel based alloys with experiments and finite element simulations. *CIRP Annals – Manufacturing Technology*, *61*(1), 547–550. https://doi.org/10.1016/j.cirp.2012.03.100

Sateesh, C., Majumder, H., Khan, A., & Kumar, S. (2020). Applicability of DLC and WC / C low friction coatings on Al 2 O 3 / TiCN mixed ceramic cutting tools for dry machining of hardened 52100 steel. *Ceramics International*, November 2019, 0–1. https://doi.org/10.1016/j.ceramint.2020.01.225

Sateesh Kumar, C., & Kumar Patel, S. (2017). Hard machining performance of PVD AlCrN coated Al2O3/TiCN ceramic inserts as a function of thin film thickness. *Ceramics International*, *43*(16), 13314–13329. https://doi.org/10.1016/j.ceramint.2017.07.030

Shalaby, M. A., Hakim, M. A. El, Abdelhameed, M. M., Krzanowski, J. E., Veldhuis, S. C., & Dosbaeva, G. K. (2014). Tribology International Wear mechanisms of several cutting tool materials in hard turning of high carbon – chromium tool steel. *Tribiology International*, *70*, 148–154. https://doi.org/10.1016/j.triboint.2013.10.011

Sui, X., Li, G., Qin, X., Yu, H., Zhou, X., Wang, K., & Wang, Q. (2015). Relationship of microstructure, mechanical properties and titanium cutting performance of TiAlN/TiAlSiN composite coated tool. *Ceramics International*, *42*(6), 7524–7532. https://doi.org/10.1016/j.ceramint.2016.01.159

Tandon, S., Jain, V. K., Kumar, P., & Rajurkar, K. P. (1990). Investigations into machining of composites. *Precision Engineering*, *12*(4), 227–238. https://doi.org/10.1016/0141-6359(90)90065-7

Thakur, A., & Gangopadhyay, S. (2016). State-of-the-art in surface integrity in machining of nickel-based super alloys. *International Journal of Machine Tools and Manufacture*, *100*, 25–54. https://doi.org/10.1016/j.ijmachtools.2015.10.001

Ucun, I., Aslantas, K., & Bedir, F. (2015). The performance Of DLC-coated and uncoated ultra-fine carbide tools in micromilling of Inconel 718. *Precision Engineering*, *41*, 135–144. https://doi.org/10.1016/j.precisioneng.2015.01.002

Wang, B., Liu, Z., & Yang, Q. (2013). Investigations of yield stress, fracture toughness, and energy distribution in high speed orthogonal cutting. *International Journal of Machine Tools and Manufacture, 73*, 1–8. https://doi.org/10.1016/j.ijmachtools.2013.05.007

Wikipedia Contributors. (2022a). *Ceramic – {Wikipedia}{,} The Free Encyclopedia.* https://en.wikipedia.org/w/index.php?title=Ceramic&oldid=1101181929

Wikipedia Contributors. (2022b). *Composite material – {Wikipedia}{,} The Free Encyclopedia.* https://en.wikipedia.org/w/index.php?title=Composite_material&oldid=1100754576

Zhvirblis, A. V. (1987). Metal-cutting. *Soviet Engineering Research, 7*(9). https://doi.org/10.4324/9781315030449-6

2 Effect of coatings on machining parameters

The previous chapter discusses the mechanism of chip formation, different metal-cutting operations, difficult-to-cut materials, and cutting tool materials. Thus, it is known that difficult-to-cut materials – due to their hardness, low thermal conductivities, and other properties – tend to generate high machining forces and cutting temperatures during material removal using the metal-cutting operation. These generated machining forces and cutting temperatures give rise to various machining challenges such as reduced cutting tool durability, and unexpected levels of surface quality for the machined surface which eventually lead to an increase in production cost and time. However, these machining challenges have been taken care of by using different cooling and lubrication techniques (Prengel et al., 2001; Sayuti et al., 2014; Settineri et al., 2008; Thakur & Gangopadhyay, 2016). The use of lubricants and coolants during machining serves different purposes. One of the most important functions of these lubricants is the reduced temperature at the cutting zone. The flowing coolant carries away a significant amount of heat by convection heat transfer resulting in reduced cutting temperatures. This reduction of temperature not only prevents the thermal deterioration of the cutting tools but also reduces the welding capacity of the flowing chips to the machined surface.

Furthermore, coolants also induce lubrication and thus reduce friction during the cutting process which eventually helps in the reduction of machining forces. The reduction in machining forces in turn results in a further reduction of cutting temperatures and leads to an increase in cutting tool durability. Although the application of coolants during machining has various advantages, the technique also suffers from disadvantages like increased production costs associated with lubrication, health hazards for workers, and sometimes thermal deterioration of the machined surface. These problems can be resolved by using minimum quantity lubrication (MQL), solid lubricants, and hard coating depositions on the cutting tools (Benedicto et al., 2017; Nee, 2015; Sateesh Kumar & Kumar Patel, 2017). In this regard, the present chapter discusses various machining challenges, solutions, and the need for thin-films in machining.

2.1 MACHINING DIFFICULTIES AND CHALLENGES

For the machining of difficult-to-cut materials, metal-cutting operations deal with different machining difficulties and challenges. As discussed earlier, the metal-cutting operation of materials like hardened steel and superalloys having low thermal conductivity generates high machining forces and cutting temperatures. This cutting environment leads to thermal deterioration and excessive abrasion of the cutting tools causing a substantial increase in tool wear rates. Furthermore, at higher cutting temperatures the surface quality of the machined surface can also deteriorate due to chip sticking, and built-up-edge formation (Halim et al., 2019; Martinez et al., 2017; Ucun et al., 2015). These machining challenges are discussed in detail in subsequent sections.

2.1.1 MACHINING DIFFICULTIES

Difficult-to-cut materials offer various beneficial properties that make them extremely durable under high temperatures and other adverse conditions. For instance, when we consider nickel-based superalloys, they can sustain harsh fatigue loading and high temperatures. The high hot hardness, chemical stability, and thermal stability of these superalloys make them extremely suitable for demanding applications in the fields of aerospace, defense, marine, etc. However, these superalloys exhibit properties like low thermal conductivity, chemical affinity, and work hardening when loaded which makes them one of the most difficult-to-cut materials (Thakur et al., 2014). On the other hand, hardened steel due to its high hardness requires larger machining forces for the removal of material in the form of chips (Kumar & Patel, 2017; Sateesh Kumar & Kumar Patel, 2017). When we consider properties like high hot hardness, the material retains high hardness even at high temperatures. However, this property is extremely beneficial as far as the usability of the material is considered under adverse conditions. On the other hand, high hot hardness results in high machining forces, and thus, higher cutting temperatures when compared to the material that softens rapidly when temperature increases. Furthermore, the lower thermal conductivity of the material causes more heat flow to the cutting tool leading to the thermal failure of the cutting tools (Jeon et al., 2014; Koyilada et al., 2016; Li et al., 2002).

In this regard, Figures 2.1 and 2.2 show the chip formation process while cutting (a) using uncoated and (b) using TiN-coated mixed ceramic cutting tool and the

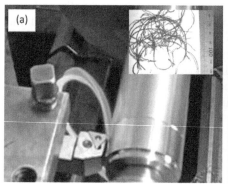

FIGURE 2.1 Chip formation while machining hardened AISI 52100 steel with (a) uncoated and (b) TiN-coated mixed ceramic cutting tools **Source:** for details see "Aslantas, K., Ucun, T. I., & çicek, A. (2012). Tool life and wear mechanism of coated and uncoated Al2O3/TiCN mixed ceramic tools in turning hardened alloy steel. Wear, 274–275, 442–451. https://doi.org/10.1016/j.wear.2011.11.010". Reprinted with permission from Elsevier

Effect of coatings on machining parameters

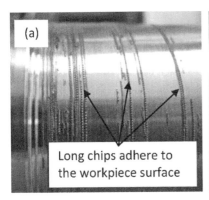

FIGURE 2.2 Quality of machined surface (a) chip adhesion and (b) chip welding to the workpiece material **Source:** for details see "Aslantas, K., Ucun, T. I., & çicek, A. (2012). Tool life and wear mechanism of coated and uncoated Al2O3/TiCN mixed ceramic tools in turning hardened alloy steel. Wear, 274–275, 442–451. https://doi.org/10.1016/j.wear.2011.11.010". Reprinted with permission from Elsevier

quality of the machined surface during the machining of hardened AISI 52100 steel (Aslantas et al., 2012). It is evident from the figures that the machining of hardened steel has resulted in high cutting temperatures visible from the red hot chips formed during the metal-cutting operation. Also, the chips formed have a dark complexion indicating burning due to high temperatures. Also, when the machined surface has been observed, two surface deterioration phenomena were observed. The long chips formed during the machining process adhered to the workpiece surface. Also, a significant amount of chips are welded to the machined surface. These phenomena occur due to high cutting temperatures and forces, providing sufficient pressure and temperature for adhesion and welding to take place.

Figure 2.3 shows the variation of cutting temperatures with cutting speed and feed rate during the machining of hardened AISI 52100 steel (Kumar & Patel, 2018a). It has been observed that the measured cutting temperatures increase with the increase of both cutting speed and feed rate. This happens because of the increase of material removal rate with the increase of cutting speed and the increase of the tool-workpiece contact region with the increase of feed rate respectively. Leaving aside discussion related to the effect of thin-films on the cutting temperatures, which will be discussed in future chapters, it is observed that machining hardened steel resulted in extremely high cutting temperatures for both coated and uncoated cutting tools. These machining outcomes like high cutting temperatures and forces in combination with extreme material properties such as high hot hardness, surface hardness, low thermal conductivity, and chip adhesion to the workpiece surface result in various machining challenges.

2.1.2 Machining challenges

The machining challenges that arise during the machining of difficult-to-cut materials due to the machining difficulties discussed above are high tool wear rates,

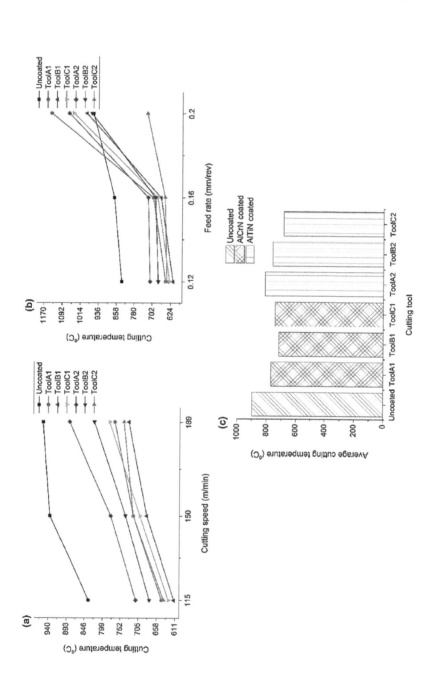

FIGURE 2.3 Variation of cutting temperatures (a) with cutting speed, (b) with feed rate, and (c) average cutting temperatures for uncoated and coated cutting tools **Source:** for details see "Kumar, C. S., & Patel, S. K. (2018). Investigations on the effect of thickness and structure of AlCr and AlTi based nitride coatings during hard machining process. Journal of Manufacturing Processes, 31, 336–347. https://doi.org/10.1016/j.jmapro.2017.11.03". Reprinted with permission from Elsevier

Effect of coatings on machining parameters

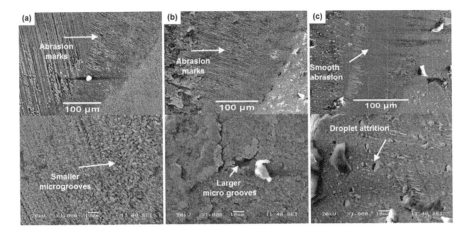

FIGURE 2.4 Wear on the rake surface of (a) uncoated, (b) AlCrN coated and (c) AlTiN coated Al$_2$O$_3$/TiCN mixed ceramic cutting tools **Source:** for details see "Kumar, C. S., & Patel, S. K. (2017). Effect of chip sliding velocity and temperature on the wear behaviour of PVD AlCrN and AlTiN coated mixed alumina cutting tools during turning of hardened steel. Surface and Coatings Technology, 334 (December 2017), 509–525. https://doi.org/10.1016/j.surfcoat.2017.12.013". Reprinted with permission from Elsevier

poor quality of the machined surface, and increased production costs and time. The high cutting forces and generated heat not only result in thermal deterioration of the cutting tools but also provide sufficient temperature and pressure conditions for the adhesion of the workpiece or chip material to the cutting tool. This adhesion may take place near the cutting edge resulting in the formation of a built-up edge (BUE) and also at the chip-tool interface causing the formation of the adhesion layer (Grzesik, 2009; Kumar & Patel, 2017; Martinez et al., 2017). In this regard, Figure 2.4 shows the wear to the rake surface of Al$_2$O$_3$/TiCN mixed ceramic cutting tools after machining for AISI 52100 steel hardened to 63 HRC hardness. The wear is characterized by abrasion, adhesion, and groove formation for the uncoated cutting tool whereas attrition in addition to abrasion and adhesion can be seen for the coated cutting tools. From the figure, it is evident that the wear during machining for difficult-to-cut materials is on the high side. Furthermore, as discussed earlier, the high cutting temperatures and machining forces prevailing result in chip adhesion to the machined surface leading to the poor surface quality of the machined surface. The elevated tool wear rates and unexpected surface integrity eventually reduce the cutting tool durability and surface quality of the machined surface, respectively. This further increases production costs and time.

2.2 AVAILABLE REMEDIES FOR MACHINING DIFFICULTIES

Various methods are employed to tackle machining challenges and difficulties. The techniques basically focus on the reduction of friction and temperature during the metal-cutting operation. The reduction in friction eventually has a cumulative effect on the metal-cutting outcomes. The reduced friction results in a reduction in machining forces which in turn reduces cutting temperatures. Thus, a significant reduction of friction during the machining operation not only results in an exponential increase in the cutting tool's durability but also improves the surface quality of the machined surface substantially (An et al., 2014; Arulkirubakaran et al., 2016; Chinchanikar & Choudhury, 2014; Engineering & Ford, 2009; Shokrani et al., 2017; Yazid et al., 2011). In this regard, different remedies to machining difficulties will be discussed one by one.

2.2.1 Cutting fluids

The cutting fluids when used during machining provide different benefits. They tend to reduce the cutting temperatures by removing a significant amount of heat during the metal-cutting operation. Further, they act as a viscous lubricant between the contacting surface thereby providing sufficient lubrication which in turn results in a reduction of friction. As discussed earlier, the reduction of friction reduces the machining forces and cutting temperatures. Thus, the cutting fluids deliver the function of a coolant as well as a lubricant. Conventionally water is used as a cutting fluid in most flood cooling systems in combination with other lubricating chemicals. However, there have been various advances in the field of cutting fluids such as using minimum quantity lubrication (MQL), the use of nano-fluids as cutting fluids, using compressed air as cutting fluids, and the implementation of different oils for their cooling and lubricating properties (Benedicto et al., 2017; Chinchanikar & Choudhury, 2015; Sayuti et al., 2014; Shokrani et al., 2017; Thakur & Gangopadhyay, 2016).

2.2.2 Solid lubricants

In addition to cutting fluids, solid lubricants like graphite and MOS_2 can also be used for providing lubrication during machining operations. Intense investigations on the effect of cutting fluids during metal-cutting operations revealed that the reduction of friction during the machining process can significantly improve the quality of the machined surface, cutting tool durability, and reduce the hazardous environmental effects of conventional cooling techniques. However, the use of solid lubricants creates a serious cleaning paradigm due to the sticking effect of the solid lubricants. However, researchers have proved that the use of solid lubricants with MQL can be very effective during machining applications (Benedicto et al., 2017; Vamsi Krishna & Nageswara Rao, 2008; Yang et al., 2013).

2.2.3 Coatings

The disadvantages of conventional cutting fluids and solid lubricants can be overcome by the use of thin-film depositions on the cutting tools. The deposited coatings on the cutting tools not only act as a thermal barrier between the tool substrate and the workpiece but also provide lubrication and wear resistance during the machining process (Kumar & Patel, 2018b; Sateesh Kumar et al., 2020; Thakur et al., 2014). Furthermore, the coatings eliminate cleaning processes arising due to the cutting fluids on the workpiece and the machine tool and also due to the sticking of solid lubricants. These coatings may have different architecture/structure, hardness, behavior, and application which will be elaborated on in Chapter 4. The wear resistance provided by the coatings on the cutting tools in addition to lubrication is an added advantage for improving cutting tool durability during the metal-cutting operation.

2.3 NEED FOR THIN-FILMS IN MACHINING

From the above discussion on machining difficulties and challenges, and possible remedies for tackling those machining challenges it is evident that the cutting fluids possess significant disadvantages such as the generation of hazardous vapors due to the interaction of fluids and the heated machined surface, tool, and chips. These vapors give rise to different environmental and health hazards for workers. On the other hand, the use of solid lubricants can solve the hazardous effects of conventional cutting fluids to some extent, but still, the problem is not completely eliminated. In addition, solid lubricants sticking to the workpiece and the machine tools calls for additional cleaning. These disadvantages can be eliminated by the use of thin-film depositions on the cutting tools. As discussed earlier, deposited coatings provide wear resistance in addition to lubrication. Furthermore, the coatings can exhibit properties like the generation of lubricious phases during the machining operation, high hardness due to nanocomposite structure, and a multilayered structure providing crack deflection to name a few (Brzezinka et al., 2019; Kuo et al., 2017; Panjan et al., 2003). Thus, the coatings provide a wide range of properties that can be beneficial during the machining of difficult-to-cut materials. In addition, the coatings can eliminate the use of cutting fluids completely leading to environment-friendly machining operations. These functions of thin-films have made them an integral part of machining applications.

REFERENCES

An, Q., Wang, C., Xu, J., Liu, P., & Chen, M. (2014). Experimental investigation on hard milling of high strength steel using PVD-AlTiN coated cemented carbide tool. *International Journal of Refractory Metals and Hard Materials*, *43*, 94–101. https://doi.org/10.1016/j.ijrmhm.2013.11.007

Arulkirubakaran, D., Senthilkumar, V., & Kumawat, V. (2016). Effect of micro-textured tools on machining of Ti–6Al–4V alloy: An experimental and numerical approach. *International Journal of Refractory Metals and Hard Materials*, *54*, 165–177. https://doi.org/10.1016/j.ijrmhm.2015.07.027

Aslantas, K., Ucun, T. I., & çicek, A. (2012). Tool life and wear mechanism of coated and uncoated Al2O3/TiCN mixed ceramic tools in turning hardened alloy steel. *Wear*, *274–275*, 442–451. https://doi.org/10.1016/j.wear.2011.11.010

Benedicto, E., Carou, D., & Rubio, E. M. (2017). Technical, economic, and environmental review of the lubrication/cooling systems used in machining processes. *Procedia Engineering*, *184*, 99–116. https://doi.org/10.1016/j.proeng.2017.04.075

Brzezinka, T., Rao, J., Paiva, J., Kohlscheen, J., Fox-Rabinovich, G., Veldhuis, S., & Endrino, J. (2019). DLC and DLC-WS2 coatings for machining of aluminium alloys. *Coatings*, *9*(3), 192. https://doi.org/10.3390/coatings9030192

Chinchanikar, S., & Choudhury, S. K. (2014). Hard turning using HiPIMS-coated carbide tools: Wear behavior under dry and minimum quantity lubrication (MQL). *Measurement: Journal of the International Measurement Confederation, 55*, 536–548. https://doi.org/10.1016/j.measurement.2014.06.002

Chinchanikar, S., & Choudhury, S. K. (2015). Machining of hardened steel – Experimental investigations, performance modeling and cooling techniques: A review. *International Journal of Machine Tools and Manufacture, 89*, 95–109. https://doi.org/10.1016/j.ijmachtools.2014.11.002

Grzesik, W. (2009). Wear development on wiper Al2O3-TiC mixed ceramic tools in hard machining of high strength steel. *Wear, 266*(9–10), 1021–1028. https://doi.org/10.1016/j.wear.2009.02.010

Halim, N. H. A., Haron, C. H. C., Ghani, J. A., & Azhar, M. F. (2019). Tool wear and chip morphology in high-speed milling of hardened Inconel 718 under dry and cryogenic CO 2 conditions. *Wear, 426–427*(January), 1683–1690. https://doi.org/10.1016/j.wear.2019.01.095

Jeon, S., Van Tyne, C. J., & Lee, H. (2014). Degradation of TiAlN coatings by the accelerated life test using pulsed laser ablation. *Ceramics International, 40*(6), 8677–8685. https://doi.org/10.1016/j.ceramint.2014.01.085

Koyilada, B., Gangopadhyay, S., & Thakur, A. (2016). Comparative evaluation of machinability characteristics of Nimonic C-263 using CVD and PVD coated tools. *Measurement, 85*, 152–163. https://doi.org/10.1016/j.measurement.2016.02.023

Kumar, C. S., & Patel, S. K. (2017). Effect of chip sliding velocity and temperature on the wear behaviour of PVD AlCrN and AlTiN coated mixed alumina cutting tools during turning of hardened steel. *Surface and Coatings Technology, 334*(December 2017), 509–525. https://doi.org/10.1016/j.surfcoat.2017.12.013

Kumar, C. S., & Patel, S. K. (2018a). Investigations on the effect of thickness and structure of AlCr and AlTi based nitride coatings during hard machining process. *Journal of Manufacturing Processes, 31*, 336–347. https://doi.org/10.1016/j.jmapro.2017.11.031

Kumar, C. S., & Patel, S. K. (2018b). Performance analysis and comparative assessment of nano-composite TiAlSiN/TiSiN/TiAlN coating in hard turning of AISI 52100 steel. *Surface and Coatings Technology, 335*(September 2017), 265–279. https://doi.org/10.1016/j.surfcoat.2017.12.048

Kuo, Y., Wang, C., & Lee, J. (2017). The microstructure and mechanical properties evaluation of CrTiAlSiN coatings : Effects of silicon content. *Thin Solid Films, 638*, 220–229. https://doi.org/10.1016/j.tsf.2017.07.058

Lei, S., Devarajan, R., Zenghu, C. (2009). A comparative study on the machining performance of textured cutting tools with lubrication. *International Journal of Mechatronics and Manufacturing Systems, 2*(4), 401–413.

Li, L., He, N., Wang, M., & Wang, Z. G. (2002). High speed cutting of Inconel 718 with coated carbide and ceramic inserts. *Journal of Materials Processing Technology, 129*(1–3), 127–130. https://doi.org/10.1016/S0924-0136(02)00590-3

Martinez, I., Tanaka, R., Yamane, Y., Sekiya, K., Yamada, K., Ishihara, T., & Furuya, S. (2017). Wear mechanism of coated tools in the turning of ductile cast iron having wide range of tensile strength. *Precision Engineering, 47*, 46–53. https://doi.org/10.1016/j.precisioneng.2016.07.003

Nee, A. Y. C. (2015). Handbook of manufacturing engineering and technology. In *HandBook of Manufacturing Engineering and Technology.* https://doi.org/10.1007/978-1-4471-4670-4

Panjan, P., Čekada, M., & Navinšek, B. (2003). A new experimental method for studying the cracking behaviour of PVD multilayer coatings. *Surface and Coatings Technology, 174–175*, 55–62. https://doi.org/10.1016/S0257-8972(03)00618-2

Prengel, H. G., Jindal, P. C., Wendt, K. H., Santhanam, A. T., Hegde, P. L., & Penich, R. M. (2001). A new class of high performance PVD coatings for carbide cutting tools. *Surface and Coatings Technology*, *139*(1), 25–34. https://doi.org/10.1016/S0257-8972(00)01080-X

Sateesh Kumar, C., & Kumar Patel, S. (2017). Hard machining performance of PVD AlCrN coated Al2O3/TiCN ceramic inserts as a function of thin film thickness. *Ceramics International*, *43*(16), 13314–13329. https://doi.org/10.1016/j.ceramint.2017.07.030

Sateesh Kumar, C., Majumder, H., Khan, A., & Patel, S. K. (2020). Applicability of DLC and WC/C low friction coatings on Al2O3/TiCN mixed ceramic cutting tools for dry machining of hardened 52100 steel. *Ceramics International*, *46*(8), 11889–11897. https://doi.org/10.1016/j.ceramint.2020.01.225

Sayuti, M., Sarhan, A. A. D., & Salem, F. (2014). Novel uses of SiO2 nano-lubrication system in hard turning process of hardened steel AISI4140 for less tool wear, surface roughness and oil consumption. *Journal of Cleaner Production*, *67*, 265–276. https://doi.org/10.1016/j.jclepro.2013.12.052

Settineri, L., Faga, M. G., & Lerga, B. (2008). Properties and performances of innovative coated tools for turning inconel. *International Journal of Machine Tools and Manufacture*, *48*(7–8), 815–823. https://doi.org/10.1016/j.ijmachtools.2007.12.007

Shokrani, A., Dhokia, V., & Newman, S. T. (2017). Hybrid cooling and lubricating technology for CNC milling of inconel 718 nickel alloy. *Procedia Manufacturing*, *11*(June), 625–632. https://doi.org/10.1016/j.promfg.2017.07.160

Thakur, A., & Gangopadhyay, S. (2016). Dry machining of nickel-based super alloy as a sustainable alternative using TiN/TiAlN coated tool. *Journal of Cleaner Production*, *129*, 256–268. https://doi.org/10.1016/j.jclepro.2016.04.074

Thakur, A., Mohanty, A., & Gangopadhyay, S. (2014). Comparative study of surface integrity aspects of Incoloy 825 during machining with uncoated and CVD multilayer coated inserts. *Applied Surface Science*, *320*, 829–837. https://doi.org/10.1016/j.apsusc.2014.09.129

Thakur, A., Mohanty, A., Gangopadhyay, S., & Maity, K. P. (2014). Tool wear and chip characteristics during dry turning of inconel 825. *Procedia Materials Science*, *5*, 2169–2177. https://doi.org/10.1016/j.mspro.2014.07.422

Ucun, I., Aslantas, K., & Bedir, F. (2015). The performance Of DLC-coated and uncoated ultra-fine carbide tools in micromilling of Inconel 718. *Precision Engineering*, *41*, 135–144. https://doi.org/10.1016/j.precisioneng.2015.01.002

Vamsi Krishna, P., & Nageswara Rao, D. (2008). Performance evaluation of solid lubricants in terms of machining parameters in turning. *International Journal of Machine Tools and Manufacture*, *48*(10), 1131–1137. https://doi.org/10.1016/j.ijmachtools.2008.01.012

Yang, J. F., Jiang, Y., Hardell, J., Prakash, B., & Fang, Q. F. (2013). Influence of service temperature on tribological characteristics of self-lubricant coatings: A review. *Frontiers of Materials Science*, *7*(1), 28–39. https://doi.org/10.1007/s11706-013-0190-z

Yazid, M. Z. A., Ibrahim, G. A., Said, A. Y. M., CheHaron, C. H., & Ghani, J. A. (2011). Surface integrity of Inconel 718 when finish turning with PVD coated carbide tool under MQL. *Procedia Engineering*, *19*, 396–401. https://doi.org/10.1016/j.proeng.2011.11.131

3 Coating technologies

In the modern age, surface engineering and advanced coatings are used right across the spectrum of manufacturing and engineering industries to enhance the surfaces of components that can be made from low-cost, lightweight, or sustainable materials made of polymers, metals, ceramics, composites, or even biological material. This allows the possibility to create cost-effective, high-performance parts with a functional surface exactly where it is required. Machining tools are a good example; the tools are made of hard steel or tungsten carbide and then coated to extend the surface capabilities (high hardness, high wear resistance, high oxidation resistance, low friction, etc.) and consequently improve the tool's lifetime. Therefore, in the following sub-sections, the main deposition technologies used to protect the surface and consequently extend the service life of machining tools is presented in detail.

3.1 PHYSICAL VAPOR DEPOSITION

The importance of this technology for the industrial sector has been increasing over the past years owing to its versatility in the deposition of films with tunable chemical composition and good quality. In the machining industry, the technology is widely used to protect the surface of tools mainly with nitrides (Baptista et al., 2018; Zhang et al., 2019).

In the middle of the eighteenth century, researchers observed that if two electrodes were placed inside of a chamber in vacuum conditions, a very thin layer of material could be deposited over the cathode, anode, and chamber walls. This phenomenon is called "sputtering". Sputtering designates thus the mechanism of ejection of material from a surface (target) by the bombardment of particles with high energy. The ejected material travels in a vacuum up to the surface to be protected. Figure 3.1 shows a schematic representation of the variants which can be distinguished in the sputtering process.

The different configurations for the sputtering machine depend on the application/work conditions from which the films will be working. The different power supplies which can be used have an extreme influence on the film's mechanical properties and morphology. On the other hand, the motion of the electrons is very important to define the sputtering efficiency. In diode magnetron sputtering configurations, the sputtering yield is very low as there is no efficient ionization of the Ar gas (Ar^+) to bombard the target surface. In a triode sputtering system there are three main electrodes with separated supply sources: (i) cathode heating voltage supply, (ii) plasma voltage supply (a voltage applied to the anode-cathode pair to maintain the plasma), and (iii) high voltage sputtering supply (applied to the sputtering target), which are used to improve the sputtering yield of the material. In the former case, with the introduction of a magnetron below the target of the material to be deposited, the motion of secondary electrons is limited to the vicinity of the target surface and the ionization fraction of Ar gas (or inert gas used to serve as plasma) exponentially increased. This contributes to the increase of the sputtered material from the targets. Depending on the configuration of the magnetrons the process can be called balanced magnetron sputtering and unbalanced magnetron sputtering, see Figure 3.2. In the case of balanced magnetron sputtering the magnetic field lines are closed and the electron cannot escape from the magnetic field. This results in a plasma that will be formed around the cathode. This means that the electron cannot escape from the magnetic field generated, creating a strong plasma near the target surface (plasma shines due to the ionization of Ar ($Ar^+ + e^-$)). For the unbalanced magnetron sputtering two magnets configuration can be used: (i) a weaker central magnet and strong magnets in the border; and (ii) a strong central magnet and weak ones in the border.

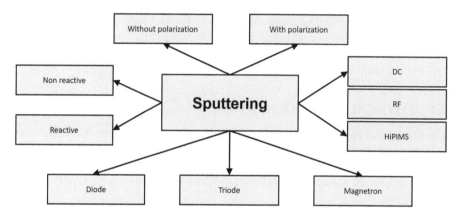

FIGURE 3.1 Variants of the sputtering technology

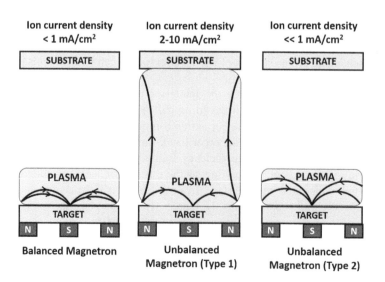

FIGURE 3.2 Balanced and unbalanced magnetron sputtering configurations

In the former case, the magnetic field lines will not be caught up by the central region, and the shape of the resultant magnetic field will extend, allowing the easy escape of electrons from the plasma toward the substrate. In the second case, the unclosed magnetic field lines are pointed toward the chamber walls and the plasma density in the substrate region is low.

3.1.1 BASICS OF THE PROCESS

As mentioned previously, the conventional sputtering process involves the ejection of atoms from the target surface due to the bombardment with energetic ions (Møller & Nielsen, 2012). An inert gas is introduced into the deposition chamber, which will ionize to form positive ions. These ions are accelerated toward the target that is negatively charged. Due to the collision of these ions with the material from the target, radiation, reflection of ions, and ejection of atoms will occur from the target material to produce the film. In the ionization process (dissociation of the inert gas), plasma is established inside the chamber and needs to be maintained for the process to occur continuously. To ensure that the plasma is maintained, a sufficiently high potential difference between electrodes must be used so that secondary electrons are ejected when the gas ions collide with the target. Indeed, enough electrons must be produced to induce the formation of ions, which in turn must be capable, on impact with the target, of ejecting secondary electrons again in similar proportion, to allow continuous ionization of the gas. For the ionization process to continue, one must also ensure that the pressure inside the chamber is not too low to promote a sufficient number of collisions between electron and gas atoms. However, for high-pressure values, the ion will suffer too many collisions, which leads to the loss of energy, thus preventing the collision with the target from having enough energy to eject secondary electrons, which impairs the maintenance of the plasma.

Since the collision has an elastic character, not all the energy is consumed in the collision. Thus, the atoms of the material ejected out after bombardment have enough kinetic energy to travel up to the substrate. As already mentioned, the pressure must not be too low, since the number of particles inside the chamber will be relatively high and consequently will suffer many collisions (gas diffusion process) during the time of flight. The average free distance, which is the average distance that a particle travels without suffering collisions, when low, leads to the thermalization of particles, i.e., loss of their initial energy in a very small distance avoiding the particles to reach the substrate. This promotes a decrease in the deposition rate. The sputtering efficiency is defined by the mass relationship of ion type and target material. Ar is the most used gas due to its low cost and availability on the market. On the other hand, their atomic mass guarantees an optimum sputtering yield for almost all the elements. However, other gases such as Xe, Kr, and Ne are also often used. The use of inert gases has the advantage of not producing compounds, although they can stay trapped in the films. A schematic representation of the sputtering process using conventional power supplies is displayed in Figure 3.3.

In the last few decades, new magnetron sputtering deposition techniques have been developed to produce highly ionized fluxes of sputtered material and, hence, to allow increased control over the energetic ion bombardment (energy and direction of

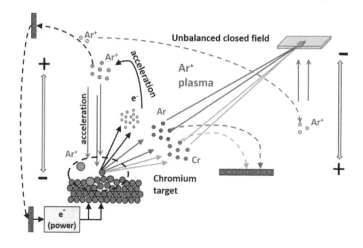

FIGURE 3.3 Schematic representation of the sputtering conventional process

the deposited species) (Fernandes et al., 2020; Oliveira et al., 2015). One of the recent developments is the HiPIMS (high power impulse magnetron sputtering technology) technology. Whilst in conventional sputtering power supplies the power applied to the targets is in the range of 0.5–10 kW (depending on the dimension of the target), in HiPIMS power supplies the power applied to the targets can be extremely high, i.e. in the range of several hundred kW. This allows for the achievement of higher plasma densities and subsequent ionization of the sputtered material. Thus, some differences in the deposition process can be highlighted as follows: (i) ionization of the material of the target occurs (the level of ionization depends on the deposition parameters, being the maximum of ionization reported to be in the range of 70–80%), (ii) metallic ions from the target, injected out can be backtracked to the target, decreasing the

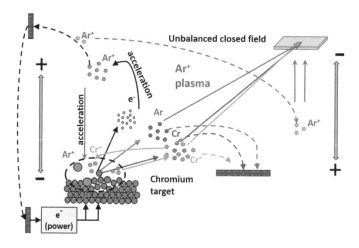

FIGURE 3.4 Schematic representation of the sputtering occurring in the HiPIMS process

deposition rate, (iii) the ions arriving to the substrate have more energy and mobility producing more compact and harder films. Figure 3.4 shows a schematic representation of the different steps happening in the HiPIMS sputtering process. The main disadvantages of the films produced with HiPIMS are the low deposition rate of the process and the induction of high levels of residual stress on the films. This deposition rate problem is being overcome by increasing the number of magnetrons in the chamber. However, recently a new HiPIMS variant, which allows the deposition during the off time of the HiPIMS pulse, was developed by an enterprise in Spain (Nano4Energy), allowing the increase of the deposition rate.

The process described before either for conventional sputtering and HiPIMS power supplies is only taking into account the deposition of material from which the target is made (single elements or composite materials). However, the material to be deposited is often produced by adding a reactive gas in the chamber to produce a compound combining the chemical elements of the target with one of the gases. This is called "reactive magnetron sputtering" (Thull & Grant, 2001). The majority of reactive-based sputtering processes are characterized by a hysteresis-like behavior, thus needing proper control of the involved parameters, e.g. the partial working pressure and reactive gas flow. The stoichiometry of the films can be thus controlled by changing the relative pressure of the inert and reactive gases. This technique is used to deposit compounds like nitrides, carbides, and oxides. The main advantages of the reactive process over the non-reactive one are (i) easier production of compounds since in the non-reactive process it is necessary to produce targets of those compounds with adequate composition (for certain materials the production of targets is very complex and/or even impossible), (ii) allows the deposition of insulator materials with DC power supplies which would not be possible using a single target of that insulator material and (iii) allows the deposition of films with gradient chemical composition, by controlling the flux of gases. Nevertheless, the poisoning of the target with the reactive gas leads to a decrease in the deposition rate. Additionally, the control of the process is often difficult. For example, in oxides, the transition from the metallic to the stoichiometric compound occurs abruptly. Thus, in those cases, intermediary compounds may be impossible to be deposited in reactive mode. In those cases, the hysteresis curve should be deeply studied and special flowmeters and plasma analysis techniques should be used to properly control the process.

3.1.2 MORPHOLOGY OF THE FILMS

In conventional sputtering processes the morphology of the films is well known to be highly influenced by the deposition pressure, deposition temperature, distance of the substrates to the targets, and the applied voltage at the substrate. Thornton (2014), developed a model which allowed the prediction of the morphology of the films as a function of the deposition pressure and deposition temperature (the temperature at the substrates). Figure 3.5 represents the influence of the deposition pressure and temperature on the morphology of the films produced. From this diagram, it can be observed that there are four types of referenced coatings: type 1, 2, T, and 3. The compactness level of the films increases from type 1 to 3 which is connected with the substrate temperature and consequently with the high mobility of the atoms. During

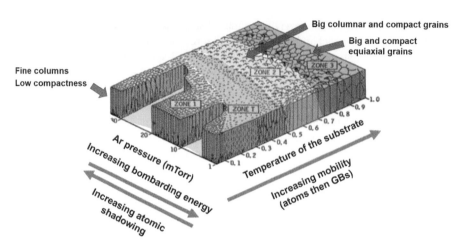

FIGURE 3.5 Thornton diagram showing the type of morphology expected as a function of the deposition pressure and deposition temperature. Diagram adapted from Thornton, 2014

the film growth, the arriving atoms at the surface will move in the surface of the growing film. If the energy of the arriving atoms to the growing surface is low, they will stay at the arriving positions producing open morphologies, such as type 1 and T. Moreover, when the deposition pressure is high, the ejected atoms from the target can lose energy motivated by the collisions (depending on the mean free path) and can reach the substrate at tangential trajectories. This favors the shadowing effect mechanism and consequently leads to the production of columnar films. Although the Thornton diagram can predict the type of morphology of conventional sputtering technologies as a function of the deposition temperature and deposition pressure, it cannot be used to predict the morphology of films produced by HiPIMS technologies since it does not account for the ionization fraction and energy of particles arriving at the substrate. It should be noted that in HiPIMS, ions with higher mobility and energy are reaching the substrate leading to more compact structures.

The bias (negative voltage) applied at the substrate has a huge impact on the morphology of the films, since it influences the nucleation and growth of the film and the mobility of the adatoms at the film surface. In conventional sputtering processes, the bias applied to the substrate attracts the Ar ions, bombarding the surface of the film. This bombardment in one way reduces the surface roughness avoiding the shadowing effect and consequently producing more compact films. On the other hand, the bombardment by Ar ions increases the mobility of the adatoms at the surface, by kinetic energy transfer, contributing to the increase of compactness. A schematic representation is shown in Figure 3.6. Bias applied to the substrate also allows for improving the adhesion of the films to the substrate. This bombardment by Ar ions is also used before depositions to clean the substrates in one step called "etching". In this step, very high voltages applied to the substrate are used. Nevertheless, some disadvantages can also be pointed out when bias is applied to the substrates: (i) excessive bias leads to the formation of films high an extremely high level of compressive residual stress, which leads to the easier delamination and or failure of the films after deposition or in service condition, (ii) under high bias and therefore

Coating technologies

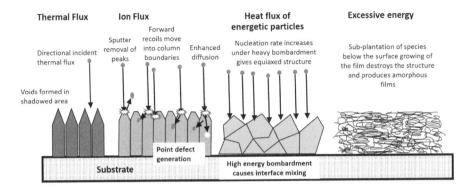

FIGURE 3.6 Change in the microstructure with ions bombardment

intense ions bombardment, the ions can be implanted in the sub-surface of the growing films producing amorphous structures undesirable in terms of mechanical properties, and (iii) under intense ions bombardment lighter elements are easily re-sputtered, being very difficult to produce films with adequate chemical composition. In the case of the HiPIMS power supply, metallic ions from the target material will be attracted by the bias of the substrate, increasing the level of compactness. Due to the different ionized species produced during the discharge (see time-averaged ion energy distributions (IEDFs) graph measured for TiN films produced with different reactive N gas, in Figure 3.7), a synchronized bias is often used to attract specific ions to the substrate and thus to tune the properties of the films.

3.2 CHEMICAL VAPOR DEPOSITION

An alternative technology to the PVD method is the deposition of coatings using chemical vapor deposition (CVD). CVD is a generic name for a set of processes that involves the deposition of solid material on the surface of a substrate during the vapor phase of a controlled chemical reaction (Ritala et al., 2009). The basic principle of the CVD method lies in the decomposition of high-purity gases in a chamber under controlled conditions, which further undergo a series of chemical and physical reactions and culminate with the deposition of a solid phase over the substrates. The CVD technique is a versatile and quick method to support film growth, enabling the generation of pure coatings with uniform thickness and controlled porosity, even on complicated or contoured surfaces. As compared to conventional PVD technology the following advantages and disadvantages can be highlighted:

Advantages

- High-quality films with less porosity and more compacts.
- Higher deposition rates.
- Stoichiometry of films is relatively easy to control.
- Ease of doping.
- Complex-shaped objects can be coated, inclusively interior holes.
- The homogeneous thickness of the films over complex surfaces.

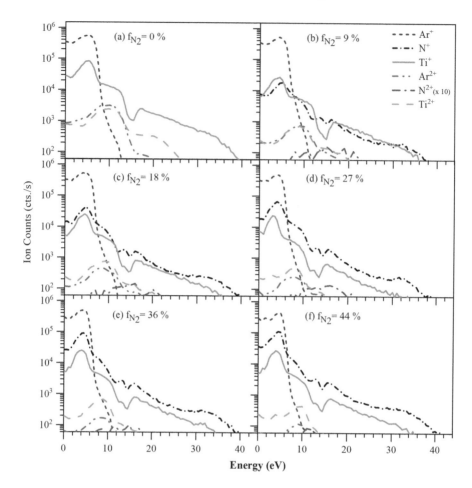

FIGURE 3.7 Time-averaged IEDFs measured for increasing N2 fraction in the discharge gas, fN2 =: a. 0%; b. 9%; c. 18%; d. 27%; e. 36%; and f. 44% for producing TiN films (Oliveira et al., 2018)

Disadvantages

- Complicated reaction kinetics.
- Use of thermal sources/contamination.
- High deposition temperatures are needed/restrict the type of substrates.
- Reactive, toxic, and corrosive gases are involved.
- Expensive reagents (metal-organic) are involved.

The CVD process can be concisely described by the following sequence of steps: (i) specific reagents and inert gases are introduced (with the controlled flow) in a reaction chamber; (ii) the gases diffuse to the surface of the substrate; (iii) the reagents are adsorbed on the surface; (iv) the adsorbed atoms react forming the film; (v) the reaction products are desorbed and removed from the chamber. A schematic representation of the CVD sequences is shown in Figure 3.8.

Coating technologies

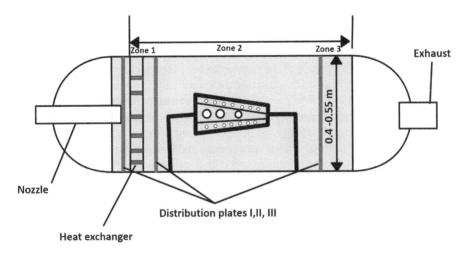

FIGURE 3.8 Schematic representation of the CVD process

The CVD process is formed by different systems. Each of the systems has a proper function as briefly described below:

- Gases system – a system that supplies the precursor gases to the reactor.

The type of gas is selected based on the final coating which is intended to be produced/deposited. The gases often used are Halides (TiCl4, TaCl5, WF6, etc.), Hydrides (SiH4, GeH4, AlH3(N(CH3)3)2, NH3, etc.), and organometallic compounds.

- Reactor (reaction chamber) – where reaction and deposition take place.

The equation below represents a general example showing a typical reaction:

$$R_1(g) + R_2(g) \rightarrow P(s) + W(g)$$

In this equation, R_1 and R_2 are the reactant gases, which will react in the gaseous phase (g). P(s) is the solid phase resulting from the reaction of the previous gases and therefore will be the solid deposited over the surfaces to be coated; W(g) is a product resulting from the reaction of the gases. Although the chemical reaction should take place only on the surface, or at least close to it ("heterogeneous reaction"), normally it can also occur in the gas phase ("homogeneous reaction"). Thus, the reaction above is a simplification of the reaction process, since it can be divided into a series of side reactions occurring at the same time in different zones of the CVD chamber. A schematic representation of those zones is plotted in Figure 3.9.

In this figure in zone A, intermediate reactions/gases may be formed if necessary for the succeeding surface reactions. In zone B, called the diffusion zone, diffusion processes in the boundary zone between the solid surfaces and the reaction gases occur. The diffusion processes vary depending on the chosen precursors and substrate, temperature, chamber pressure, carrier gas flow rate, quantity and ratio of source materials, and source-substrate distance for the CVD process. In zone C, atoms and molecules adsorb and desorb in the surface zone. In zone D there occurs the growth and nucleation of the coating. Finally, in zone E diffusion processes between the outermost region of the coating and the grown coating occurs.

As mentioned previously, the structure and morphology of the produced films by CVD are a function of several parameters. The parameters that influence the structure of the film are the temperature and deposition rate. Figure 3.10 represents the relation between the temperature and deposition rate with the type of structure that can be obtained. In summary, high temperature and low deposition rates favor epitaxial growth, whilst low temperature and high deposition rates favor the formation of amorphous structures. Intermediary temperature and deposition rate conditions

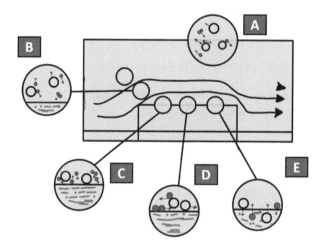

FIGURE 3.9 Schematic representation of the different reactions

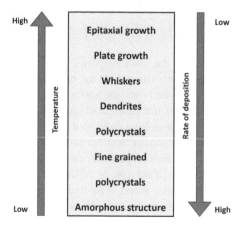

FIGURE 3.10 General structure of the films as function of the deposition rate and temperature produced by CVD technology

Coating technologies

may allow the formation of plate growth, whiskers, dendrites, polycrystals, or fine-grained polycrystals.

The CVD reactors can be divided into two categories: cold and hot wall reactors. This classification is summarized in Figure 3.11 and depends on factors such as operating pressure, the temperature of the reactor walls, and the use of heating sources. In the cold wall reactors the substrates are warmed up directly using for instance induction heating, whilst, in hot wall reactors, the reactor is warmed up using a resistive heating system. In the former, contaminations coming from the reactor wall will be avoided; however, is it difficult to maintain a homogeneous temperature over the parts. Hot wall reactors allow the reach of homogeneous temperatures on the parts.

Different materials can be deposited by CVD as metals, metal alloys, carbides, nitrides, borides, oxides, and intermetallic compounds. As mentioned previously, the gases should be properly selected to ensure adequate material over the parts to be protected. Below are some examples of CVD reactions taking place inside the reactors to produce the protective materials:

$CH4_{(g)} \rightarrow C_{(s)} + 2H2_{(g)}$, this reaction is used to produce carbon, graphite, and diamond.

$$TiCl_{4(g)} + 2BCl_{3(g)} + 5H_{2(g)} \rightarrow TiB_{2(s)} + 10HCl_{(g)}$$

$$SiH_{4(g)} + O_{2(g)} \rightarrow SiO_{2(s)} + 2H_{2(g)}$$

$$NbCl_{5(g)} + \frac{5}{2}H_{2(g)} + \frac{1}{2}H_{2(g)} \rightarrow NbN_{x(s)} + 5HCl_{(g)}$$

Sources of energy for the reaction

The required energy for the reaction can be provided by different sources such as resistive heating (e.g. "diffusion" ovens), radiative heating (e.g. halogen lamps),

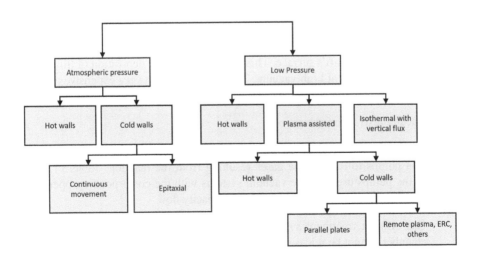

FIGURE 3.11 Reactor types

radiofrequency/induction heating, using plasma, and heating using a laser. Among the previous sources, the ones based on thermal energy are the most used (the so-called thermal CVD). The types of CVD are: atmospheric pressure chemical vapor deposition (APCVD), low-pressure chemical vapor deposition (LPCVD), metal-organic chemical vapor deposition (MOCVD), plasma assisted chemical vapor deposition (PACVD) or plasma enhanced chemical vapor deposition (PECVD), and laser chemical vapor deposition (LCVD).

The APCVD was the first process to be developed. It is simple, uses cheap equipment, and allows high deposition rates. However, since it is susceptible to reactions in the gas phase, it needs high gas flow rates. The process does not allow for the production of homogeneous coverage of irregular surfaces with steps and contamination by particles can occur.

The LPCVD produces better films than APCVD reactors in terms of uniformity of the deposited film, step covering, and particulate contamination. However, the equipment is more complex, high temperatures are involved, and the deposition rate is lower.

The MOCVD is a particular case of LPCVD, made at moderate pressures (primary vacuum. It is widely used in the growth of epitaxial grain. This process is characterized by having an organometallic reagent that disintegrates easily in the presence of moderate temperatures.

PACVD is probably the most common type of CVD. The process uses plasma to provide some of the energy for the reaction to occur rather than having to heat the substrate to such a high degree to drive these reactions. The gas ionization allows a reactivity normally obtained at high temperatures, i.e. the possibility of using low temperatures of deposition. The process temperature is lower than in the APCVD and LPCVD allowing the deposition of films in low melting point substrates. However, generally, the films obtained are not stoichiometric and the sub-products from the reaction, especially hydrogen, oxygen, and nitrogen, can be incorporated into the film. An excess of these contaminants can cause the film to crack and eventually make them delaminate over consecutive thermal cycles.

In the LCVD process, the energy of the focused laser beam is absorbed by reagent gases, leading to the decomposition of gas molecules and the formation of the material over the part's surface. This method allows the deposition of films with high purity and works at high vacuum. This method is also very expensive and the deposition rates are very low.

- Vacuum/extraction system

The CVD systems have pumps to produce a vacuum inside the chamber reactor and extraction pumps to remove the gases resulting from the reaction. Table 3.1 summarizes the pumps used for the vacuum systems with the corresponding operating temperature range and maximum extraction capacity flow.

- Gas effluent treatment system – The gases extracted from the reaction chamber cannot be released into the atmosphere without prior treatment.
- Control equipment of the system –

TABLE 3.1
Vacuum pump classifications and some properties

Vacuum pump	Operating conditions (Torr)	Maximum capacities (l/s)
Water jet pump	760 to 15–25	500
water ring pump	760 to 75 – 1st stage; 125 to 20 – 2nd stage	1 to 2500
steam ejector	760 to 75 – 1st stage; 125 to 20 – 2nd stage; 30 to 2.5 – 3rd stage; 5 to 0.03 – 4th stage	100000
Oil-sealed rotatory pump	760 to 2×10^{-2} – 1st stage; 760 to 5×10^{-3} – 2nd stage	0.25 to 500
Roots pump	10 to 10^{-3}	50 to 35000
Vacuum diffusion pump	10^{-2} to 10^{-9}	95000
Oil vapor booster pump	1 to 10^{-4}	23000
Sputtering ion pump	10^{-2} to 10^{-11}	7 000
Radial field pump	10^{-4} to 10^{-11}	400 to 800
Titanium sublimation pump	10^{-3} to 10^{-11}	Thousands
Sportion pump	760 to 10^{-2} – 1st stage; 760 to 10^{-5} – multi-stage	1000
Molecular pump	10^{-1} to 10^{-10}	10000
Cryopump	10^{-3} to 10^{-10}	Million

Pressure, flow, temperature, time, and security alarms are very important parameters to control during depositions. Specific equipment is used to control those magnitudes.

3.3 HYBRID DEPOSITION TECHNIQUES

Hybrid deposition technologies combine more than one deposition process. When the deposition of coatings over coating tolls is to be undertaken, four mainly hybrid deposition processes can be considered:

- Conventional PVD + HiPIMS.
- Conventional PVD + Arc (cathodic arc discharge).
- HiPIMS + Arc.
- Conventional PVD + PECVD.

The arrangement of the technologies in a chamber allows one way to increase the deposition rate of the material being deposited and in the other, to tune the mechanical properties of the coatings. These technologies can be either working separately or together in the same deposition chamber. In some coatings, one of the processes is used to produce the inter/gradient layers to improve the adhesion of the films to the substrate, and the other to produce the final film. Multilayer arrangement of the coatings with the different processes is often produced to reach the so-called superlattice effect.

REFERENCES

Baptista, A., Silva, F. J. G., Porteiro, J., Míguez, J. L., Pinto, G., & Fernandes, L. (2018). On the physical vapour deposition (PVD): Evolution of magnetron sputtering processes for industrial applications. *Procedia Manufacturing, 17*, 746–757. https://doi.org/10.1016/j.promfg.2018.10.125

Fernandes, F., Calderon V. S., Ferreira, P. J., Cavaleiro, A., & Oliveira, J. C. (2020). Low peak power deposition regime in HiPIMS: Deposition of hard and dense nanocomposite Ti-Si-N films by DOMS without the need of energetic bombardment. *Surface and Coatings Technology, 397*, 125996. https://doi.org/10.1016/j.surfcoat.2020.125996

Møller, P., & Nielsen, L. P. (2012). *Advanced Surface Technology: Volume 01* (Issue vol. 1). https://books.google.pt/books?id=wF0RzQEACAAJ

Oliveira, J. C., Fernandes, F., Ferreira, F., & Cavaleiro, A. (2015). Tailoring the nanostructure of Ti-Si-N thin films by HiPIMS in deep oscillation magnetron sputtering (DOMS) mode. *Surface and Coatings Technology, 264*. https://doi.org/10.1016/j.surfcoat.2014.12.065

Oliveira, J. C., Fernandes, F., Serra, R., & Cavaleiro, A. (2018). On the role of the energetic species in TiN thin film growth by reactive deep oscillation magnetron sputtering in Ar/N2. *Thin Solid Films, 645*. https://doi.org/10.1016/j.tsf.2017.10.052

Ritala, M., Niinisto, J., Krumdieck, S., Chalker, P., Aspinall, H., Pemble, M. E., Gladfelter, W. L., Leese, B., Fischer, R. A., Parala, H., Kanjolia, R., Dupuis, R. D., Alexandrov, S. E., Irvine, S. J. C., Palgrave, R., & Parkin, I. P. (2009). *Chemical Vapour Deposition* (A. C. Jones & M. L. Hitchman (eds.)). The Royal Society of Chemistry. https://doi.org/10.1039/9781847558794

Thornton, J. A. (2014). Influence of apparatus geometry and deposition conditions on the structure and topography of thick sputtered coatings. *Journal of Vacuum Science and Technology, 666*(1974), 2–7. https://doi.org/10.1116/1.1312732

Thull, R., & Grant, D. (2001). Physical and chemical vapor deposition and plasma-assisted techniques for coating titanium. In *Titanium in Medicine: Material Science, Surface Science, Engineering, Biological Responses and Medical Applications* (pp. 283–341). Springer, Berlin Heidelberg. https://doi.org/10.1007/978-3-642-56486-4_10

Zhang, Q., Wu, Z., Xu, Y. X., Wang, Q., Chen, L., & Kim, K. H. (2019). Improving the mechanical and anti-wear properties of AlTiN coatings by the hybrid arc and sputtering deposition. *Surface and Coatings Technology, 378*, 125022. https://doi.org/10.1016/j.surfcoat.2019.125022

4 Classification of coatings

4.1 GENERATION OF COATINGS

With the development of advanced materials with superior properties and due to the strong competition between businesses, the machining industry has pressed to manufacture parts at lower cost, while ensuring that they operate efficiently, are friendly to the environment, and meet the safety requirements. These advanced materials have gained considerable importance for the construction of safe and more energy-efficient vehicles in the aerospace and automotive industries. Market analysis reports from 2021 to 2027 (*Reportsanddata*, n.d.) show a continuous incremental increase in the use of superalloys, as displayed in Figure 4.1. For instance, the demand for titanium alloys has been steadily increasing in different industries because of their excellent strength-to-weight ratio and electrochemical compatibility. Titanium alloys have outstanding physical-mechanical properties, but they are difficult-to-machine materials because of their high chemical reactivity, poor heat conductivity, and low modulus of elasticity, which lead to lower production rates and increased tool wear. The unusual serrated chip formation and low thermal conductivity tend to increase the temperature on the edge and rake face of the cutting tools leading to its premature failure.

Machined parts today are not only defined by the most demanding rigor, delivery, and quality but increasingly are made from difficult-to-machine materials that many manufacturers have never faced before, not to mention the growing number of cases where composites are replacing metals. In this domain, machining equipments are constantly evolving and certainly getting faster and more accurate; nevertheless, the cutting tools are not evolving in the same way. In this industry the increase of the material removal rates is a very important factor to decrease machining times, producing parts at a lower cost without compromising the surface quality of the parts with the main goal to make a profit, as much as possible. It could be thought that an obvious way to increase the material removal rates would be the increase of the cutting parameters. However, the increase of cutting parameters generates very high friction, extremely high cutting forces, and strong adhesive interaction with the tool's material which increases the temperature on the contact leading to both the premature failure of the tool and poor surface finishing of the parts. Therefore, this is not an adequate solution for the machining industries. The effect of cutting speed on the machining cost is presented in Figure 4.2 (adapted from Inspektor and Salvador, 2014). One effective way to decrease the temperature of the contact is the use of cutting fluids. Cutting fluids are indeed often used to decrease the temperature of the cutting contact and provide lubrication; however, they are harmful to the environment and hard to disposal. Thus there is a pressing demand from the EU regulators to remove them from the market. Different strategies are currently being applied to

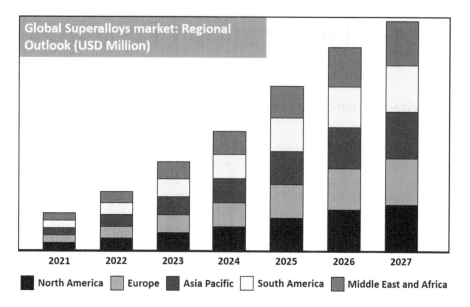

FIGURE 4.1 Past, present, and future global market size of super-alloys (period of 2012–2024) (Reportsanddata, n.d.)

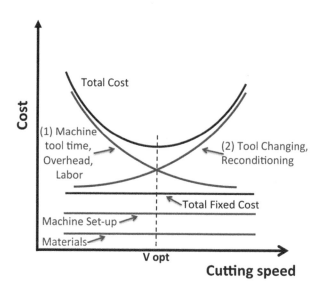

FIGURE 4.2 Effect of the machining speed on the machining costs. Adapted from Inspektor & Salvador, 2014)

improve and extend the service life of machining tools and the quality finishing of the parts such as:

- Using high-conductive cutting tools and tool holders.
- Application of coolant.
- Hybrid machining.
- Evaluation of the temperature, stresses, and vibration and in situ correct the machining parameters.
- Application of hard protective coatings.

Among the previous solutions the use of coatings gained particular interest. The cost of tooling is typically in the range of 2–5% of the total manufacturing cost, nevertheless, its real capacity lies in the increase of productivity by increasing the metal removal rate and reducing machining costs.

Different types of coatings are being used to protect the surface of the tools using the processes described in Chapter 3 and other not presented in this book. The first coating system applied to improve the service life of tools was the TiN in the year of 1980 (Wolfe et al., 1986). CrN hard coating was also one of the first generations of protective coatings used for cutting. However, due to their low oxidation resistance, ~600 °C (Chim et al., 2009) and 700 °C (Lin et al., 2008) for TiN and CrN, respectively, the properties of those systems were further improved by allowing it with a third element. In particular, Ti(CN) coatings, the first PVD multilayer TiN/Ti(CN), and monolithic TiAlN (1998) coatings, with steep improvements in the oxidation resistance, thermal stability, and hot hardness were developed. Other Ti-, Cr-, and Al-based coatings were developed and deposited as monolayer and multilayer configurations, in particular the AlTiN and AlCrN coatings. Later in the mid-1990s nanocomposite coatings with superior mechanical properties were introduced, essentially the TiSiN system (Veprek & Reiprich, 1995). Si was added to the general nitride coatings to produce this kind of structure which allowed the improvement of the mechanical properties, thermal stability, and oxidation resistance. Carbide-based coatings have been also deposited over the tools; however, due to their lower fracture toughness and oxidation resistance their application was limited. The new generation of coatings is being developed by adding other elements and playing with the chemical composition and architecture of coatings and using more advanced deposition technologies such as HiPIMS. The research is focused on the development of coatings with multiple properties such as high hardness, fracture toughness, low friction, thermal stability, oxidation resistance, etc.

4.2 CLASSIFICATION BASED ON THE COATING MATERIAL

It is known that the performance of the coatings is significantly dependent on their chemical composition (Bertóti, 2002). The coatings used to protect the surface of the machining tools can be sub-divided into five groups: nitrides, carbides, borides, oxides, and carbo-nitrides. One more than others allowed successfully to improve the machining performance of different base materials. Their selection is made depending on the operation conditions, material to be machined, thermal stability,

TABLE 4.1

Classification of coating based on coating material and some examples of the coating systems

Type	Nitrides	Carbides	Borides	Oxides	carbo-nitrides
Material	TiN, CrN, TiAlN, CrAlN, etc	NbC, CrC, TiC, WC, etc	TiB_2, ZrB_2, B_4C etc	Al_2O_3, Cr_2O_3, TiO, Zr_2O_3, etc	TiCN, NbCN, etc

tribological performance, and on the affinity of the coating to the base material to be machined (coatings which can easily establish chemical bonds with the elements of the substrate should not be used). However, among the available solutions nitrides have been particularly developed and used due to their better combination of properties. Table 4.1 summarizes the classification of coating based on coating material and some examples of the coating systems.

4.3 CLASSIFICATION BASED ON COATING PROPERTIES

The coatings used to extend the service life of machining tools can be classified mainly as hard, soft, and low friction coatings. The hardness of the coatings is a very important property to take into account, although it also depends on the fracture toughness, wear mechanisms, etc. Figure 4.3 displays the resistance to crater wear as a function of the resistance to flank wear for different hard coatings. The most popular systems among the hard coatings are the c-Ti1–xAlxN and c-Cr1–xAlxN coatings, with AlN content (x) close to its metastable cubic solubility (solubility limit ~ 0.7). With this composition hardness values higher than 30 and 37 GPA, respectively,

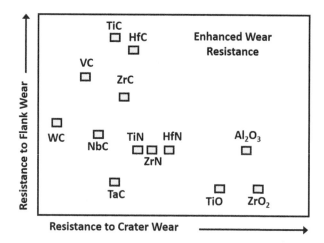

FIGURE 4.3 Wear resistance of different hard coatings, displayed as a function of the resistance to crater wear and resistance to flank wear. Adapted from Inspektor and Salvador, 2014)

can be reached. However, in the former coating a concentration of Al close to 50 at.% is preferable due to the better compromise of properties. This coating system had great success in machining since it allowed excellent heat insulation between chips and tools. taC carbon films (often designated as diamond-like carbon films – DLCs) with a strong level of sp3 bonds can also be considered hard coatings since hardness values in the range of 30–50Gpa can be reached. Although hard, this coating system can also be considered as low friction coating, since low friction coefficient values are provided by this coating. Soft coatings are less hard than hard coatings; however, fracture toughness is likely improved. Such type of coatings, normally metallic or non-stoichiometric coatings with high amounts of metallic elements, can be applied over the cutting tools to machine under low cutting speeds or moderate cutting conditions; nevertheless, they are not adequate for dry and high-speed cutting conditions, especially when super-alloys are considered. It is also common in both hard and soft coatings to have low-friction properties. This property depends on the ability of elements to establish low friction tribolyaers.

4.4 CLASSIFICATION BASED ON COATING ARCHITECTURE/DESIGN

The design of the coatings is a very important factor to take into account during their development. Coatings with suitable adhesion to the substrate, mechanical, thermal, oxidation, tribological properties, and chemical inertness are required. Their architecture design can be dived into four categories: monolayer, multilayer, double layer, and nanocomposite coatings, as illustrated in Figure 4.4.

Independently of the configuration adhesion and gradient layers are often used to improve the adhesion of coatings to the substrate, accommodate the differences in hardness between the substrate and coatings, and to attenuate stress fields normally established in the interface substrate/coating. In the monolayer coatings the working layer with a thickness which normally varies between 1 and 3 μm is deposited over the substrate or the optimized inter-gradient layers. The composition of this layer is normally constant but in specific cases a gradient composition varying the concentration of specific elements can be deposited. As mentioned in sub-section 4.1 the first generation of monolayer coatings were TiN and CrN. Those coatings are chemically

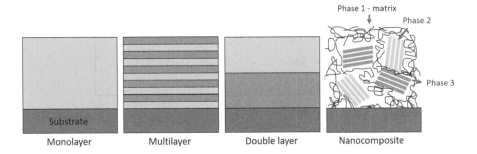

FIGURE 4.4 Typical coating architectures/design

stable; however, they are susceptible to abrasive wear. Their use is recommended for more ductile materials, such as low-carbon steels and aluminum, submitted to low cutting forces. In the monolayer coatings when a crack initiates at the top, they propagate through the cross-section leading to the delamination of the coating. To avoid such crack propagation multilayered coatings were developed. Multilayer coatings are arrangements of two layers with different chemical compositions repeated several times within a specific period. Depending on the period thickness of the layers the so-called superlative effect can be reached, i.e. extremely high hardness values can be achieved. The evolution of hardness as a function of the period thickness of layers displays a typical behavior, where the high hardness is reached for a period thickness in the range of 2 to 10 nm. Then the hardness decreases and remains fairly constant as shown in Figure 4.5, for the multilayer TiN/VN coating system (Stueber et al., 2009). The superlative effect is caused by the blocking dislocations motion at the layer interfaces, due to differences in the shear moduli of the materials of the individual layers and by coherency strain causing periodical strain-stress fields in the case of lattice-mismatched multilayered films. The constant hardness values reached with increasing period thickness represent the weighted average of the hardness of the individual layers.

Nanocomposites being composed of nanocrystalline phases, either thermodynamically stable or metastable and amorphous phases are considered a new generation of advanced multifunctional materials. These special multiphase coatings are subject to intensive R&D efforts with regard to the design of new protective

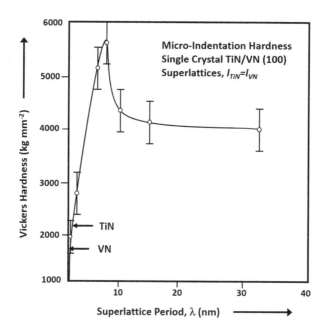

FIGURE 4.5 Hardness of the multilayer TiN/VN system as function of the period thickness. Adapted from Helmersson et al., 1987)

coatings with superior properties and performance since the first reports on superhard composite coatings were published almost one decade ago (Veprek & Reiprich, 1995). Their predicted unique physical and functional properties are attributed to the particular nanophase arrangement and the dominating role of interfaces (Gleiter, 2000), and are the main driving force for the research and development activities worldwide on these materials. The selection of materials to be combined in such a multiphase structure should refer to their phase relations in the corresponding phase diagram. The resulting microstructure and property profile of a nanocomposite coating is clearly determined by the deposition process parameters and by the kinetics of the deposition and growth process. The properties of a nanocomposite structure can be completely different by either fine-tuning the ratio of the volume fraction of the nanocrystalline and of the amorphous phases or by fine-tuning the crystallite sizes. Consequently, the properties of nanocomposite coatings are dominated by interphases and interfaces. The formation of nanocomposite structures is connected with a segregation of the one-phase to grain boundaries of the second phase, and this effect is responsible for stopping grain growth and blocking the dislocation motions when a specific size of grains and thickness of the matrix is reached, thus leading to high hardness values.

REFERENCES

Bertóti, I. (2002). Characterization of nitride coatings by XPS. *Surface and Coatings Technology*, *151–152*, 194–203. https://doi.org/10.1016/S0257-8972(01)01619-X

Chim, Y. C., Ding, X. Z., Zeng, X. T., & Zhang, S. (2009). Oxidation resistance of TiN, CrN, TiAlN and CrAlN coatings deposited by lateral rotating cathode arc. *Thin Solid Films*, *517*(17), 4845–4849. https://doi.org/10.1016/j.tsf.2009.03.038

Gleiter, H. (2000). Nanostructured materials: Basic concepts and microstructure. *Acta Materialia*, *48*(1), 1–29. https://doi.org/10.1016/S1359-6454(99)00285-2

Helmersson, U., Todorova, S., Barnett, S. A., Sundgren, J. E., Markert, L. C., & Greene, J. E. (1987). Growth of single-crystal TiN/VN strained-layer superlattices with extremely high mechanical hardness. *Journal of Applied Physics*, *62*(2), 481–484. https://doi.org/10.1063/1.339770

Inspektor, A., & Salvador, P. A. (2014). Architecture of PVD coatings for metalcutting applications: A review. *Surface and Coatings Technology*, *257*, 138–153. https://doi.org/10.1016/j.surfcoat.2014.08.068

Lin, J., Mishra, B., Moore, J. J., & Sproul, W. D. (2008). A study of the oxidation behavior of CrN and CrAlN thin films in air using DSC and TGA analyses. *Surface and Coatings Technology*, *202*(14), 3272–3283. https://doi.org/10.1016/j.surfcoat.2007.11.037

reportsanddata. (n.d.). Retrieved September 14, 2022, from https://www.reportsanddata.com/report-detail/superalloys-market

Stueber, M., Holleck, H., Leiste, H., Seemann, K., Ulrich, S., & Ziebert, C. (2009). Concepts for the design of advanced nanoscale PVD multilayer protective thin films. *Journal of Alloys and Compounds*, *483*(1), 321–333. https://doi.org/10.1016/j.jallcom.2008.08.133

Veprek, S., & Reiprich, S. (1995). A concept for the design of novel superhard coatings. *Thin Solid Films*, *268*(1), 64–71. https://doi.org/10.1016/0040-6090(95)06695-0

Wolfe, G. J., Petrosky, C. J., & Quinto, D. T. (1986). The role of hard coatings in carbide milling tools. *Journal of Vacuum Science & Technology A: Vacuum, Surfaces, and Films*, *4*(6), 2747–2754. https://doi.org/10.1116/1.573673

5 Application of coatings for machining

Thin-films may have different properties depending on the coating deposition technology, coating architecture/structure, and chemical composition (Hovsepian et al., 2005; Kumar & Patel, 2018; Prengel et al., 2001; Vereschaka et al., 2014). Depending on these properties the coatings can be classified into hard, low-friction, hybrid, and duplex coating systems. The hard coatings possess high hardness that results in superior wear resistance. These coatings in addition to providing wear resistance, can reduce friction, and act as a thermal barrier (Endrino et al., 2006; Kumar et al., 2020; Sateesh Kumar et al., 2020). Furthermore, the low-friction coatings are often softer but produce lubricious phases like V_2O_5, WO_3, graphite, etc. (Banerji et al., 2014; Hovsepian et al., 2005; Sateesh Kumar et al., 2020) during the machining operation. These low-friction phases cause a significant reduction of friction during the machining operation thereby resulting in the reduction of machining forces. These beneficial properties increase the cutting tool's durability and improve the quality of the machined surface (Fukui et al., 2004; Sateesh Kumar et al., 2020; Vandevelde et al., 1999). On the other hand there are different hybrid and duplex coating combinations whose performance has been investigated during machining (Brzezinka et al., 2019; Kumar & Patel, 2019; Xing et al., 2014). In this regard, the present chapter will discuss the application of different coating systems during the machining of difficult-to-cut materials.

5.1 HARD COATINGS

Hard coatings have proved to be very effective in improving the cutting tool's durability and generating acceptable surface integrity of the machined surface. As far as the machining of difficult-to-cut materials is considered, hard coatings like TiAlSiN, TiAlN, AlCrN, and AlTiN can be very effective in improving the machining performance of cutting tools. In this regard, Figure 5.1 shows the variation of tool life for uncoated and TiN-coated Al_2O_3–TiCN mixed inserts during the machining of hardened 52200 steel at 63 HRC hardness. The cutting tests were carried out under a dry cutting environment. It has been reported that the tool life of the cutting tools reduces with the increase in cutting speed. Furthermore, the coated cutting tools resulted in a significant increase in the tool life of the mixed ceramic inserts under all conditions. The TiN coating prevented fracture and chipping damage to the cutting tool. In addition, the machining with the coated cutting tool reduced chip sticking to the workpiece and ensured that the surface quality of the machined surface is maintained. Machining with an uncoated tool caused a decrease in chip up-curl radius due to the formation of crater wear. This reduction

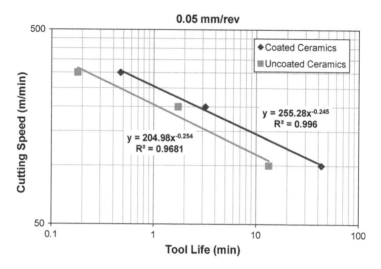

FIGURE 5.1 Tool life variation with cutting speed for uncoated and coated mixed ceramic cutting tools **Source**: For details see "Aslantas, K., Ucun, T. I., & çicek, A. (2012). Tool life and wear mechanism of coated and uncoated Al2O3/TiCN mixed ceramic tools in turning hardened alloy steel. Wear, 274–275, 442–451. https://doi.org/10.1016/j.wear.2011.11.010". Reprinted with permission from Elsevier

in chip curl radius can be attributed to the higher temperatures at the chip-tool interface causing the thermal bimetallic effect between the upper and lower side of the chips. This causes accumulation and sticking of the chips to the workpiece (Aslantas et al., 2012). Thus, tool wear behavior significantly impacts the surface integrity of the machined surface.

As per the earlier discussion, it is evident that the surface integrity of the machined surface is significantly impacted by the tool wear behavior of the cutting tools. Figure 5.2 shows the comparison of the surface morphology of the machined surface of Inconel 825 superalloy produced while machining with uncoated and CVD-coated cutting tools. Close observation revealed that the machined surface is characterized by feed marks, smeared material, redeposited material, and debris for both uncoated and coated cutting tools. These observations disclose the failure of coated tools to produce any improvement in the surface integrity of the machined surface. The high temperature at the chip-tool interface and low thermal conductivity of the Al_2O_3 coating was attributed to this behavior of coated cutting tools (Thakur et al., 2014). Thus, from the contradictory discussion regarding the effect of the coatings on the surface integrity of the machined surface, it can be concluded that the degree of wear resistance offered by the coating, chemical affinity, and other thermo-mechanical properties are highly significant when related to the surface integrity with the performance of coated cutting tools. However, if the thin-film depositions result in the reduction of friction and machining forces, and an increase of wear resistance of the cutting tools, the surface integrity, in general, would be better in comparison to that corresponding to the uncoated cutting tool.

Application of coatings for machining

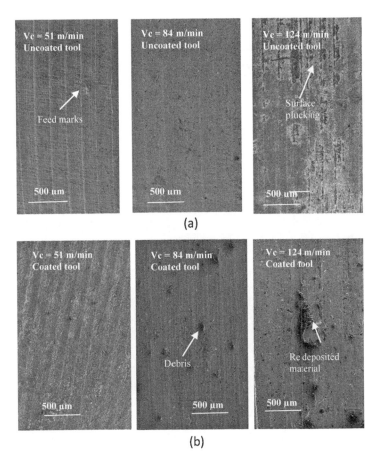

FIGURE 5.2 FESEM images showing the surface morphology of the machined surface while machining with (a) uncoated and (b) coated cutting tools **Source**: For details see "Thakur, A., Mohanty, A., & Gangopadhyay, S. (2014). Comparative study of surface integrity aspects of Incoloy 825 during machining with uncoated and CVD multilayer coated inserts. Applied Surface Science, 320, 829–837. https://doi.org/10.1016/j.apsusc.2014.09.129". Reprinted with permission from Elsevier

It has been studied by various researchers that the deposition of hard coatings on the cutting tools results in significant improvement in the machining performance (Choudhury & Chinchanikar, 2017; Dogra et al., 2012; Hood et al., 2013; Keunecke et al., 2010). The effect of thin-films on surface integrity can be elaborated by studying the surface and sub-surface regions of the machined surface as shown in Figure 5.3. From the figure it is evident that the cutting temperature significantly impacts the formation of the white layer. Furthermore, machining with coated tool resulted in the formation of a white layer only at the highest cutting speed of 124 m/min whereas the white layer was formed while machining with an uncoated cutting tool at all cutting speeds.

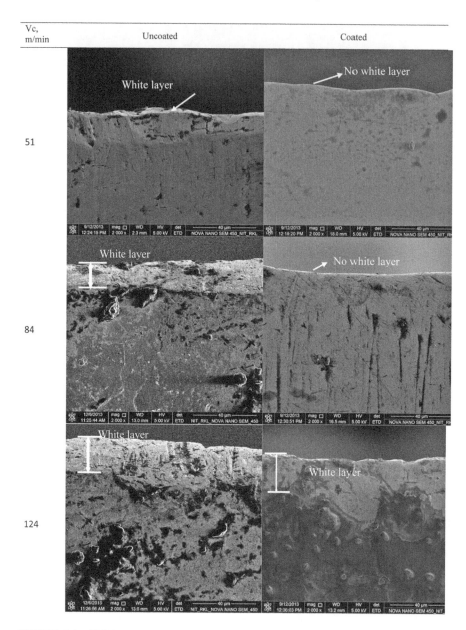

FIGURE 5.3 FESEM images showing white layer formation at different cutting speeds while machining with uncoated and coated cutting tools **Source**: For details see "Thakur, A., Mohanty, A., & Gangopadhyay, S. (2014). Comparative study of surface integrity aspects of Incoloy 825 during machining with uncoated and CVD multilayer coated inserts. Applied Surface Science, 320, 829–837. https://doi.org/10.1016/j.apsusc.2014.09.129". Reprinted with permission from Elsevier

5.2 LOW-FRICTION COATINGS

Low-friction coatings are the thin-film that diffuses out specific elements, to form a tribolayer, which by their own properties or by the combination of other elements provides low friction properties. These low-friction phases can be very beneficial in the reduction of friction and tool wear if generated during the metal-cutting operation (Al-rjoub et al., 2022). The friction reduction has a cumulative impact on the metal-cutting process in the form of the reduction of machining forces, cutting temperatures, tool wear, and better surface quality of the machined surface. However, we have already discussed how the reduction of tool wear and the chemical affinity of the coatings affect the surface integrity of the machined surface, and thus, in this section, we will discuss the effect of low-friction coatings on tool wear and cutting temperatures. In this regard, Figures 5.4 and 5.5 show the variation in cutting temperatures with cutting speed and feed rate respectively (Sateesh et al., 2020). The presented results explore the performance of DLC and WC/C low-friction coatings during the hard turning of AISI 52100 steel. From the figures, it is evident that the low-friction coatings held in the reduction of cutting temperatures under all conditions.

The DLC coating produces graphite during the machining operation whereas the WC/C coating results in the formation of WO_3. Graphite and WO_3 are both known to reduce friction and act as a low-friction phase during the machining operation (Banerji et al., 2014; Sateesh et al., 2020; Tallant et al., 1995). These low-friction phases generated during the machining process have led to the reduction of friction which in turn helps in the decrease of cutting temperatures. Further, when the effect of low-friction coatings on tool wear as shown in Figure 5.6 has been studied, a

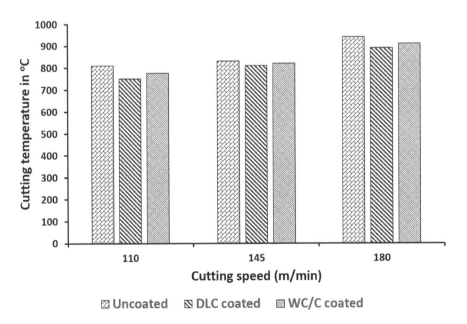

FIGURE 5.4 Variation of cutting temperature with cutting speed for uncoated, DLC-coated and WC/C-coated cutting tools

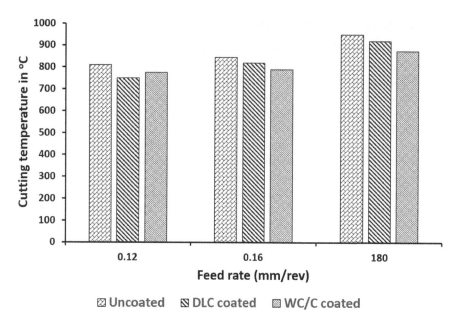

FIGURE 5.5 Variation of cutting temperature with feed rate for uncoated, DLC-coated, and WC/C-coated cutting tools

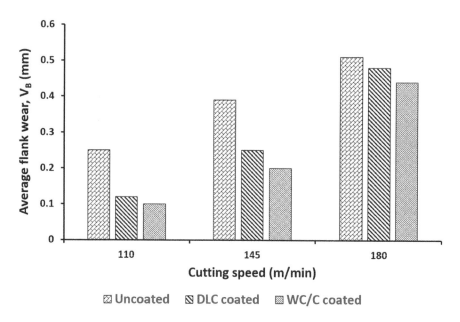

FIGURE 5.6 Variation of average flank wear V_B with cutting speed for uncoated, DLC-coated, and WC/C-coated cutting tools

significant reduction of average flank wear was observed at lower cutting speeds. But at a higher cutting speed of 180 m/min, the reduction of flank wear was not as prominent as it was in the earlier cases. This occurred due to severe coating delamination and coating stacking at higher cutting speeds due to the generation of high cutting temperatures (Sateesh et al., 2020).

5.3 HYBRID COATINGS

A hybrid coating system can be defined as a combination of hard and soft coatings. Soft coatings are basically in the form of low-friction coatings. The system is formed to get the benefit of hot hardness and wear resistance properties of hard coatings and at the same time generate lubricious phases due to the presence of low-friction coatings. In the earlier sections, it has been discussed that the deposition of both hard and low-friction coatings deposited on the cutting tools results in the reduction of tool wear, improvement of surface integrity of the machined surface, and decrease of cutting forces and temperatures. The performance of the hard coatings was attributed to their hardness and wear resistance properties in addition to their lower thermal conductivity which provided a thermal barrier effect. On the other hand, the performance of low-friction coatings was attributed to the generation of lubricious phases during the machining operation causing the reduction of friction. However, these low-friction coatings suffer from low coating/substrate adhesion strength leading to coating delamination and poor performance under adverse machining conditions.

One such example has been shown in Figure 5.7 which illustrates the variation of flank wear with cutting length, and SEM morphology showing tool wear on the cutting inserts coated with AlTiN, AlTiN/MoN, and AlTiN/NbN coatings generated during the machining of Inconel 718 superalloy at 40 m/min and 60 m/min cutting speeds respectively (Biksa et al., 2010). From the images, it is evident that when MoN and NbN are used with AlTiN to form nano-multilayers, they help in improving the tool life of the cutting tools. Further, at higher cutting speeds the generation of lubricious oxide tribo-films helps in the reduction of friction and thus, the tool wear which is reflected through the increase in tool life.

5.4 DUPLEX COATINGS

In addition to hybrid coating systems, duplex coatings that are formed by two different layers of coatings basically of similar nature (e.g., a combination of hard and hard, or soft and soft coatings) are also being tested for machining of difficult-to-cut materials. These separate layers may have the advantages of two separate coatings. For instance, using two hard coatings like TiAlSiN and TiAlN can give the benefit of the nanocomposite structure of Si_3N_4 and TiAlN of TiAlSiN and the toughness of TiAlN when compared to the nanocomposite hard coating. In this regard, Figure 5.8 shows the variation of tool life (cutting time) with cutting speed for TiN+AlTiN duplex and other modified coating systems (Settineri et al., 2008). It is evident from the results that the duplex system has accounted for a considerable increase in the tool life for the cutting tools during the machining of the Inconel 718 superalloy. Further, the duplex system accounted for lower flank wear (see Figure 5.9) at the same

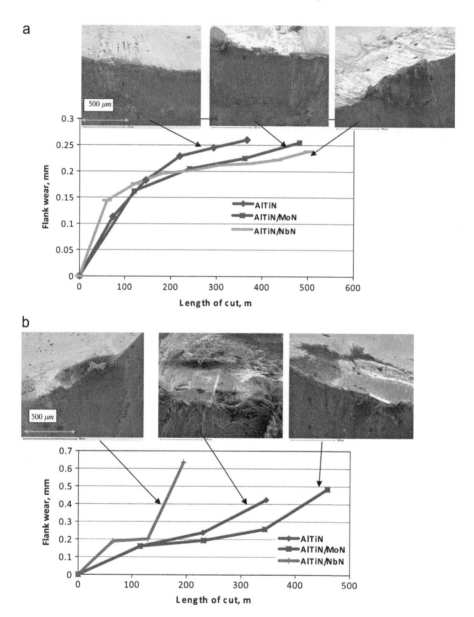

FIGURE 5.7 Flank wear vs cutting length, and SEM surface morphology showing tool wear for cutting inserts coated with AlTiN, AlTiN/MoN, and AlTiN/NbN coatings during machining of Inconel 718 at (a) 40 m/min and (b) 60 m/min cutting speed **Source**: For details see "Biksa, A., Yamamoto, K., Dosbaeva, G., Veldhuis, S. C., Fox-Rabinovich, G. S., Elfizy, A., Wagg, T., & Shuster, L. S. (2010). Wear behavior of adaptive nano-multilayered AlTiN/MexN PVD coatings during machining of aerospace alloys. Tribology International, 43(8), 1491–1499. https://doi.org/10.1016/j.triboint.2010.02.008". Reprinted with permission from Elsevier

Application of coatings for machining

cutting times when compared to the uncoated tool (Settineri et al., 2008). However, it has also been reported that the formation of a duplex layer by combining TiAlSiN and TiAlN during the machining of hardened AISI 52100 steel has not given superior results when compared to the TiAlSiN coating. However, the duplex coating accounted for a significant improvement in the machining performance in comparison to the uncoated mixed ceramic cutting tool (Kumar & Patel, 2019). Thus, from the above discussion, it is evident that the duplex coating system may or may not be beneficial when compared to a single coating system of hard or soft coatings, and depends upon the interactive properties between the coatings under consideration. Nevertheless, the duplex coatings on the cutting tools account for better machining performance when compared to the uncoated tools.

5.5 OTHER COMBINATIONS

In the previous sections, a discussion about the application of the hard, low-friction, hybrid, and duplex coatings systems during the machining of difficult-to-cut materials has been discussed. However, researchers have tried various other combinations which may work for improving the machining performance of the cutting tools (Prengel et al., 2001; Settineri et al., 2008; Wang et al., 2021). The possible combinations can be a duplex coating formed by a grouping of hard and soft coatings like TiAlN (hard) and WC/C (comparatively soft) respectively. Also, a three-layer (or more) coating system to match the hardness of the substrate can be a possible combination. Thus, in general, the coatings can be combined to take advantage of the resulting combination. The combination can be formed by depositing two hard coatings like TiN and AlTiN one over the other and then depositing a soft coating like MoS_2 to take advantage of the chemical stability of TiN coating and the high service temperature of AlTiN coating (see Figure 5.8). The MoS_2 coating can generate lubricious phases to reduce friction during the machining process. The machining performance of this modified coating system resulted in a significant reduction of tool wear (see Figure 5.9) and an increase in tool life during the machining of the Inconel

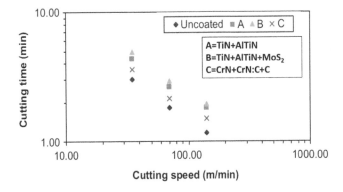

FIGURE 5.8 Variation of cutting time with cutting speed for uncoated and coated cutting tools during machining of Inconel 718 superalloy

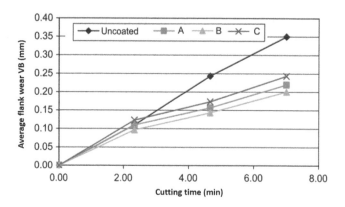

FIGURE 5.9 Variation of average flank wear with cutting time for uncoated and coated cutting tools during machining of Inconel 718 superalloy

718 superalloy. Moreover, this modified combination of the coating system with the involvement of MoS2 soft low-friction layer outperformed the duplex coating of TiN+AlTiN and another modified three-layered coating system of CrN+CrN:C+C.

From all the discussions on low-friction coatings, it can be concluded that low-friction coatings can be very beneficial in reducing friction and when used in combination with hard coatings, they can be more advantageous in terms of reducing tool wear and improving the surface integrity of the machined surface. One such example is shown in Figure 5.10 which shows SEM micrographs of the tool wear on the micro-end mills generated during the machining of Inconel 718 superalloy (Ucun et al., 2013). From the images, it is evident that chip sticking near the cutting edge is a common phenomenon for the uncoated, and AlCrN, TiAlN+AlCrN coated cutting tools. However, the AlTiN, DLC, and TiAlN+WC/C exhibited sufficient prevention against the formation of built-up-edge. The low thermal conductivity, ductile behavior, and chemical affinity of Inconel 718 resulted in chip sticking. But the higher thermal stability of AlTiN coating and the self-lubricating behavior of DLC and WC/C coatings resulted in the prevention of chip sticking.

5.6 LIMITATIONS OF THIN-FILMS FOR MACHINING OF DIFFICULT-TO-CUT-MATERIALS

In this chapter, the application of different coating systems for the machining of difficult-to-cut materials has been discussed. From the discussion, it is evident that there are certain limitations of thin-films during metal-cutting operations. Although the hard coatings provided significant wear resistance they should be selected judiciously taking into account the chemical affinity with the workpiece to be machined. For instance, AlTiN coating outperformed AlCrN coating in terms of built-up-edge formation due to the chemical affinity of Inconel 718 superalloy. Further, a modified hybrid coating CrN+ CrN:C+C exhibited poor performance when compared to other modified coating systems.

Application of coatings for machining

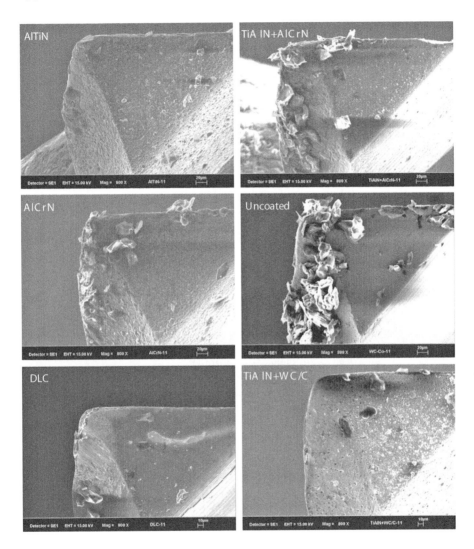

FIGURE 5.10 SEM micrographs showing wear on the micro-end mills during machining of Inconel 718 superalloy **Source**: For details see "Ucun, irfan, Aslantas, K., & Bedir, F. (2013). An experimental investigation of the effect of coating material on tool wear in micro milling of Inconel 718 super alloy. Wear, 300(1–2), 8–19. https://doi.org/10.1016/j.wear.2013.01.103". Reprinted with permission from Elsevier

The low-friction coatings helped in the reduction of friction during the machining operation but mostly they suffer from disadvantages like low hardness and coating delamination at high temperatures which are of concern during the machining of materials like nickel-based superalloys and hardened steels which have high associated cutting temperatures. Further, very hard coatings like TiAlSiN suffer from low adhesion to the substrate and thus, additional adhesion layers of coatings have to be deposited for satisfactory performance of these coatings. To conclude, the coating system should be selected in connection to the material to be machined taking into consideration the chemical affinity, thermal conductivity, hardness, grain structure, and high-temperature properties of the workpiece material.

REFERENCES

Al-rjoub, A., Bin, T., Cavaleiro, A., & Fernandes, F. (2022). The influence of V addition on the structure, mechanical properties, and oxidation behaviour of TiAlSiN coatings deposited by DC magnetron sputtering. *Journal of Materials Research and Technology, 20*, 2444–2453. https://doi.org/10.1016/j.jmrt.2022.08.009

Aslantas, K., Ucun, T. I., & çicek, A. (2012). Tool life and wear mechanism of coated and uncoated Al2O3/TiCN mixed ceramic tools in turning hardened alloy steel. *Wear, 274–275*, 442–451. https://doi.org/10.1016/j.wear.2011.11.010

Banerji, A., Bhowmick, S., & Alpas, A. T. (2014). High temperature tribological behavior of W containing diamond-like carbon (DLC) coating against titanium alloys. *Surface and Coatings Technology, 241*, 93–104. https://doi.org/10.1016/j.surfcoat.2013.10.075

Biksa, A., Yamamoto, K., Dosbaeva, G., Veldhuis, S. C., Fox-Rabinovich, G. S., Elfizy, A., Wagg, T., & Shuster, L. S. (2010). Wear behavior of adaptive nano-multilayered AlTiN/MexN PVD coatings during machining of aerospace alloys. *Tribology International, 43*(8), 1491–1499. https://doi.org/10.1016/j.triboint.2010.02.008

Brzezinka, T., Rao, J., Paiva, J., Kohlscheen, J., Fox-Rabinovich, G., Veldhuis, S., & Endrino, J. (2019). DLC and DLC-WS2 coatings for machining of aluminium alloys. *Coatings, 9*(3), 192. https://doi.org/10.3390/coatings9030192

Choudhury, S. K., & Chinchanikar, S. (2017). *1.3 Finish Machining of Hardened Steel A2 - Hashmi, MSJ BT - Comprehensive Materials Finishing* (pp. 47–92). Elsevier. https://doi.org/10.1016/B978-0-12-803581-8.09149-9

Dogra, M., Sharma, V., Sachdeva, A., & Suri, N. M. (2012). Tool life and surface integrity issues in continuous and interrupted finish hard turning with coated carbide and CBN tools. *Proceedings of the Institution of Mechanical Engineers, Part B: Journal of Engineering Manufacture, 226*(3), 431–444. https://doi.org/10.1177/0954405411418589

Endrino, J. L., Fox-Rabinovich, G. S., & Gey, C. (2006). Hard AlTiN, AlCrN PVD coatings for machining of austenitic stainless steel. *Surface and Coatings Technology, 200*(24), 6840–6845. https://doi.org/10.1016/j.surfcoat.2005.10.030

Fukui, H., Okida, J., Omori, N., Moriguchi, H., & Tsuda, K. (2004). Cutting performance of DLC coated tools in dry machining aluminum alloys. *Surface and Coatings Technology, 187*(1), 70–76. https://doi.org/10.1016/j.surfcoat.2004.01.014

Hood, R., Aspinwall, D. K., Sage, C., & Voice, W. (2013). High speed ball nose end milling of γ-TiAl alloys. *Intermetallics, 32*, 284–291. https://doi.org/10.1016/j.intermet.2012.09.011

Hovsepian, P. E., Lewis, D. B., Luo, Q., Münz, W. D., Mayrhofer, P. H., Mitterer, C., Zhou, Z., & Rainforth, W. M. (2005). TiAlN based nanoscale multilayer coatings designed to adapt their tribological properties at elevated temperatures. *Thin Solid Films, 485*(1–2), 160–168. https://doi.org/10.1016/j.tsf.2005.03.048

Keunecke, M., Stein, C., Bewilogua, K., Koelker, W., Kassel, D., & den Berg, H. van. (2010). Modified TiAlN coatings prepared by d.c. pulsed magnetron sputtering. *Surface and Coatings Technology, 205*(5), 1273–1278. https://doi.org/10.1016/j.surfcoat.2010.09.023

Kumar, C. S., & Patel, S. K. (2018). Performance analysis and comparative assessment of nano-composite TiAlSiN/TiSiN/TiAlN coating in hard turning of AISI 52100 steel. *Surface and Coatings Technology, 335*(September 2017), 265–279. https://doi.org/10.1016/j.surfcoat.2017.12.048

Kumar, C. S., & Patel, S. K. (2019). Effect of duplex nanostructured TiAlSiN/TiSiN/TiAlN-TiAlN and TiAlN-TiAlSiN/TiSiN/TiAlN coatings on the hard turning performance of Al 2 O 3 -TiCN ceramic cutting tools. *Wear, 418–419*(June 2018), 226–240. https://doi.org/10.1016/j.wear.2018.11.013

Kumar, C. S., Zeman, P., & Polcar, T. (2020). A 2D finite element approach for predicting the machining performance of nanolayered TiAlCrN coating on WC-Co cutting tool during dry turning of AISI 1045 steel. *Ceramics International, 46*(16), 25073–25088. https://doi.org/10.1016/j.ceramint.2020.06.294

Prengel, H. G., Jindal, P. C., Wendt, K. H., Santhanam, A. T., Hegde, P. L., & Penich, R. M. (2001). A new class of high performance PVD coatings for carbide cutting tools. *Surface and Coatings Technology, 139*(1), 25–34. https://doi.org/10.1016/S0257-8972(00)01080-X

Sateesh, C., Majumder, H., Khan, A., & Kumar, S. (2020). Applicability of DLC and WC / C low friction coatings on Al 2 O 3 / TiCN mixed ceramic cutting tools for dry machining of hardened 52100 steel. *Ceramics International, November 2019*, 0–1. https://doi.org/10.1016/j.ceramint.2020.01.225

Sateesh Kumar, C., Majumder, H., Khan, A., & Patel, S. K. (2020). Applicability of DLC and WC/C low friction coatings on Al2O3/TiCN mixed ceramic cutting tools for dry machining of hardened 52100 steel. *Ceramics International, 46*(8), 11889–11897. https://doi.org/10.1016/j.ceramint.2020.01.225

Settineri, L., Faga, M. G., & Lerga, B. (2008). Properties and performances of innovative coated tools for turning inconel. *International Journal of Machine Tools and Manufacture, 48*(7–8), 815–823. https://doi.org/10.1016/j.ijmachtools.2007.12.007

Tallant, D. R., Parmeter, J. E., Siegal, M. P., & Simpson, R. L. (1995). The thermal stability of diamond-like carbon. *Diamond and Related Materials, 4*(3), 191–199. https://doi.org/10.1016/0925-9635(94)00243-6

Thakur, A., Mohanty, A., & Gangopadhyay, S. (2014). Comparative study of surface integrity aspects of Incoloy 825 during machining with uncoated and CVD multilayer coated inserts. *Applied Surface Science, 320*, 829–837. https://doi.org/10.1016/j.apsusc.2014.09.129

Ucun, I., Aslantas, K., & Bedir, F. (2013). An experimental investigation of the effect of coating material on tool wear in micro milling of Inconel 718 super alloy. *Wear, 300*(1–2), 8–19. https://doi.org/10.1016/j.wear.2013.01.103

Vandevelde, T. C. S., Vandierendonck, K., Van Stappen, M., Du Mong, W., & Perremans, P. (1999). Cutting applications of DLC, hard carbon and diamond films. *Surface and Coatings Technology, 113*(1–2), 80–85. https://doi.org/10.1016/S0257-8972(98)00831-7

Vereschaka, A. A., Grigoriev, S. N., Vereschaka, A. S., Popov, A. Y., & Batako, A. D. (2014). Nano-scale multilayered composite coatings for cutting tools operating under heavy cutting conditions. *Procedia CIRP, 14*, 239–244. https://doi.org/10.1016/j.procir.2014.03.070

Wang, X., Yuan, X., Gong, D., Cheng, X., & Li, K. (2021). Optical properties and thermal stability of AlCrON-based multilayer solar selective absorbing coating for high temperature applications. *Journal of Materials Research and Technology, 15*, 6162–6174. https://doi.org/10.1016/j.jmrt.2021.11.068

Xing, Y., Deng, J., Li, S., Yue, H., Meng, R., & Gao, P. (2014). Cutting performance and wear characteristics of Al2O3/TiC ceramic cutting tools with WS2/Zr soft-coatings and nano-textures in dry cutting. *Wear, 318*(1–2), 12–26. https://doi.org/10.1016/j.wear.2014.06.001

6 Effect of coatings on machining parameters

The desirable functions of thin-films during the machining operation are to act as a thermal barrier between the tool and the workpiece, act as a lubricant by reducing friction, and provide wear resistance by increasing the surface hardness (Sateesh Kumar et al., 2020). Furthermore, coatings themselves have various governing factors such as coating thickness, coating architecture/structure, and deposition techniques that directly control the thin-film characteristics like coating/substrate adhesion strength, surface morphology, hardness, thermal, and chemical stability of the coating. These characteristics during machining play a vital role in defining the performance of coated cutting tools. Thus, the coating deposition on the cutting tools can significantly influence the output machining parameters such as machining forces, cutting temperatures, and tool wear (Kong et al., 2015). The output machining parameters are a defining factor for the tool life of the cutting tools. Thus, the coatings help in improving the tool life of cutting tools by delivering the desirable functions discussed above. However, although thin-film depositions are used to eliminate any further use of lubrication, there are certain cases where lubrication becomes necessary when machining difficult-to-cut materials like hardened steels and super alloys in adverse machining conditions like high cutting speeds and feed rates. Thus, in this chapter the effect of thin-film thickness, coating architecture/structure, and lubrication while machining with coated tools will be discussed in detail.

6.1 EFFECT OF COATING THICKNESS

The coating thickness is an important factor that affects the machining performance of cutting tools (Posti & Nieminen, 1989; Sargade et al., 2011). The increase in coating thickness affects the coating/substrate adhesion strength, cross-sectional morphology, and also microhardness of the coatings. Thus, these changes in the thin-film characteristics will significantly impact the machining performance of the cutting tools. In the present context when only difficult-to-cut materials are under consideration, the hard machining tests performed using Al_2O_3/TiCN mixed ceramic cutting tools coated with monolayered AlCrN and multilayered AlTiN coatings (Kumar & Patel, 2017; Sateesh Kumar & Kumar Patel, 2017) are considered. In both cases the coating thickness varied from 2 to 3 µm.

6.1.1 Coating characterization

As mentioned earlier, the thin-film thickness has a significant impact on the thin-film characteristics. Table 6.1 shows the measured values of surface roughness and microhardness of monolayered AlCrN and multilayered AlTiN coatings deposited

TABLE 6.1

Measured values of surface roughness, microhardness, and coating/substrate adhesion strength for monolayered AlCrN and multilayered AlTiN coatings (Kumar & Patel, 2017; Sateesh Kumar & Kumar Patel, 2017)

Coating	Coating thickness (μm)	Surface roughness, R_a (μm)	Microhardness (HV_{300})	Coating/substrate adhesion strength (N)
Monolayered AlCrN	2	0.21	3047	66
	3	0.23	3415	97
	4	0.22	3542	81
Multilayered AlTiN	2	0.21	3047	65
	3	0.21	3210	78
	4	0.21	3360	112

on the Al_2O_3/TiCN mixed ceramic substrate. It can be clearly seen that the thickness has negligible influence on the surface roughness of the coatings. However, the microhardness and the coating/substrate adhesion strength have been significantly affected by the change in thin-film thickness. The microhardness increases with the increase of thickness for both AlCrN and AlTiN coatings which apparently shows that the influence of the substrate hardness on the microhardness of the coatings reduces with the increase of the coating thickness. The variation of coating/substrate adhesion strength for the AlCrN coating is rather linear which shows higher coating thickness is beneficial for monolayered structures. However, interface residual stresses in thin-films increase with the increase in coating thickness (Qin et al., 2009) which will lead to coating delamination. On the other hand, the multilayered AlTiN coatings exhibited an increase and then a decrease in coating/substrate adhesion strength when the coating thickness changed showing higher suitability of 4 μm thickness for the coating. Thus, in this case thicker multilayer coatings will have higher resistance to delamination. Inversely it was reported that the thicker monolayered DLC coatings showed superior resistance to delamination (Qin et al., 2009). Thus, it can be concluded that the increase in residual stresses with coating thickness is a general phenomenon but coating delamination depends mostly on the coating/substrate adhesion strength.

Later, when the cross-sectional morphologies as shown in Figures 6.1 and 6.2 of the coatings were observed, the recognizable relation between the coating/substrate adhesion strength and interface could be seen for the AlTiN coating whereas AlCrN coating showed no visible changes as far as the coating/substrate interface is considered. For instance, the monolayered showed no visible interface irregularities like cracks with any coating thickness. However, for the multilayered AlTiN coating, with a 4 μm coating thickness, interface cracks are clearly visible. From the above results it is evident that the coating thickness has a noticeable impact on microhardness and the coating/substrate adhesion strength.

Effect of coatings on machining parameters

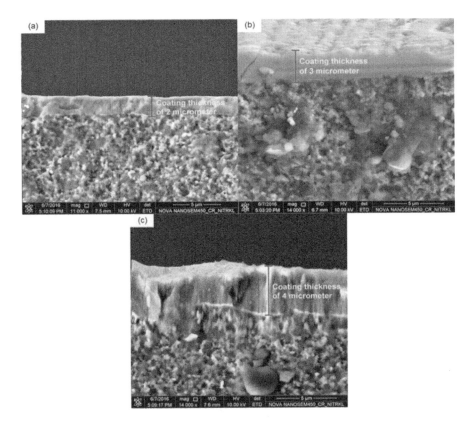

FIGURE 6.1 SEM micrographs showing cross-sectional morphologies of monolayered AlCrN coating at different thin film thicknesses **Source:** For details see: "Sateesh Kumar, C., & Kumar Patel, S. (2017). Hard machining performance of PVD AlCrN coated Al2O3/TiCN ceramic inserts as a function of thin film thickness. Ceramics International, 43(16), 13314–13329. https://doi.org/10.1016/j.ceramint.2017.07.030". Reprinted with permission from Elsevier

6.1.2 Cutting forces and temperatures

The authors also studied the variation of cutting forces and temperatures as a function of coating thickness. Figures 6.3 and 6.4 show the variation of cutting forces and temperatures with the coating thickness for AlCrN and AlTiN coatings respectively. It has been reported that the measured forces and the temperatures don't follow a particular pattern when variation in coating thickness is considered. On the other hand, they closely follow the coating/substrate adhesion strength that is reported in Table 6.1 with higher coating/substrate adhesion strength accounting for lower forces and cutting temperatures and vice versa. Thus, it is evident that the coating/substrate adhesion strength is the most dominant factor influencing the performance of cutting tools.

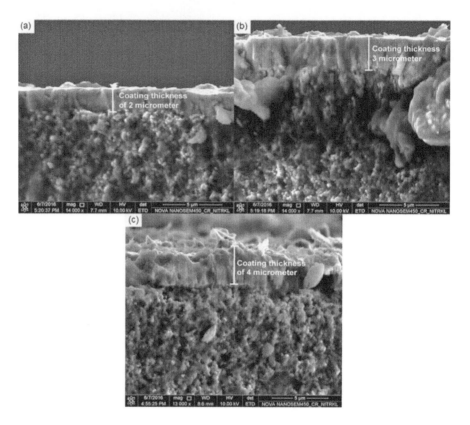

FIGURE 6.2 SEM micrographs showing cross-sectional morphologies of multilayered AlTiN coating at different thin film thicknesses **Source:** For details see: "Kumar, C. S., & Patel, S. K. (2017). Experimental and numerical investigations on the effect of varying AlTiN coating thickness on hard machining performance of Al$_2$O$_3$-TiCN mixed ceramic inserts. Surface & Coatings Technology, 309, 266–281. https://doi.org/10.1016/j.surfcoat.2016.11.080". Reprinted with permission from Elsevier

6.1.3 Chip morphology

Machining with coated cutting tools has a significant impact on chip morphology (Chinchanikar & Choudhury, 2015; Jiang et al., 2013; Thakur et al., 2015). However, not many studies are available that consider the effect of coating thickness on the chip morphology. On the other hand, when we consider only difficult-to-cut materials, again the work from Kumar and Patel (Kumar & Patel, 2017; Sateesh Kumar & Kumar Patel, 2017) can be referred to. They report that the machining with uncoated Al2O3/TiCN mixed ceramic cutting tools resulted in the formation of helical chips but the deposition of thin-films AlCrN and AlTiN led to the formation of continuous snarled chips. Also, a reduction of serration was observed while machining with coated cutting tools was directly linked to the lower coefficient of friction offered by the coatings. The chip thickness exhibited a specific trend for both AlCrN and

Effect of coatings on machining parameters

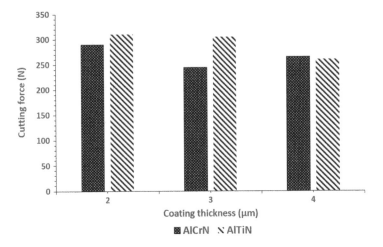

FIGURE 6.3 Variation of cutting force in N with coating thickness for monolayered AlCrN and multilayered AlTiN-coated Al_2O_3/TiCN cutting tool (Kumar & Patel, 2017; Sateesh Kumar & Kumar Patel, 2017)

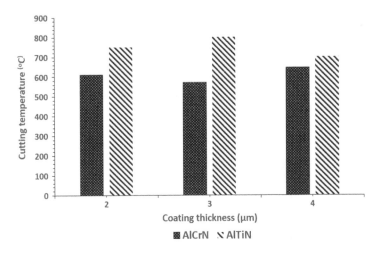

FIGURE 6.4 Variation of cutting temperature in °C with coating thickness for monolayered AlCrN and multilayered AlTiN-coated Al_2O_3/TiCN cutting tool (Kumar & Patel, 2017; Sateesh Kumar & Kumar Patel, 2017)

AlTiN coatings. The chip thickness increased with an increase in coating thickness for AlCrN coating whereas for AlTiN coating the trend was quite the opposite, with coating thickness increasing with an increase in the thin-film thickness. When combined with the results of the cutting forces and temperature, an estimation of the performance of these cutting tools can be made as a function of the thin-film thickness (Figure 6.5).

6.1.4 Tool wear

Tool wear of the cutting tools can also be affected by coating thickness. Figure 6.6 shows the variation of flank wear with coating thickness for AlCrN- and AlTiN-coated cutting tools. Also, the flank wear can be closely related to the coating/substrate adhesion strength as reported in Table 6.1. For the AlCrN coating, the tool wear was very high for 2 μm coating thickness whereas it reduced drastically when the coating thickness increased to 3 μm. On the other hand, the AlTiN coating showed a

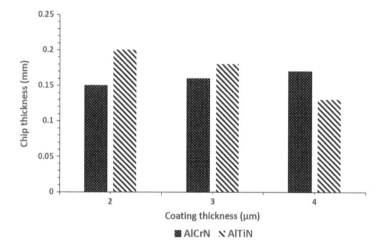

FIGURE 6.5 Variation of chip thickness in mm with coating thickness for monolayered AlCrN and multilayered AlTiN-coated Al_2O_3/TiCN cutting tool (Kumar & Patel, 2017; Sateesh Kumar & Kumar Patel, 2017)

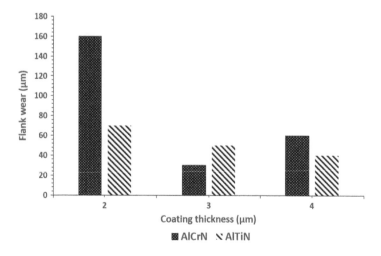

FIGURE 6.6 Variation of flank wear in μm with coating thickness for monolayered AlCrN and multilayered AlTiN-coated Al_2O_3/TiCN cutting tool (Kumar & Patel, 2017; Sateesh Kumar & Kumar Patel, 2017)

significant reduction of tool wear at all thin-film thicknesses. However, the tool wear reduced with an increase in the coating thickness. When the results of the tool wear are seen in relation to the coating/substrate adhesion strength, it can be concluded that the higher adhesion strength accounted for lower tool wear and vice versa.

6.1.5 CONCLUSIONS

This section deals with the effect of coating thickness on the performance of coated cutting tools. The performance of monolayered AlCrN and multilayered AlTiN coating are taken as references. It can be seen that the coating thickness significantly influences the thin-film characteristics that are reflected clearly in the performance of the coatings. Furthermore, monolayer and multilayered architecture also significantly impact the way the coatings behave during machining when the coating thickness is changed. Thus, the effect of architecture/structure will be elaborated on in the next section.

6.2 EFFECT OF COATING ARCHITECTURE/STRUCTURE

The structure of the coating – with monolayer, bilayer, or multilayer architecture – influences not only the coating properties like hardness, surface morphology, wear resistance, crack propagation, and oxidation resistance but can also affect the interface properties like coating/substrate adhesion strength (Kumar & Patel, 2018; Toparli et al., 2007; Vereschaka et al., 2017). The coating strain i.e., the resistance to cracking, is higher for the multilayered coatings. Also, it has been reported that the penetration of cracks from the surface to the substrate is almost straight with no noticeable deviation due to the coating topography or surface morphology for monolayered coatings like TiN and NbN (see Figure 6.7 (a)). Whereas for multilayered coatings like TiN/NbN, the movement of the crack to the substrate is rather deflected

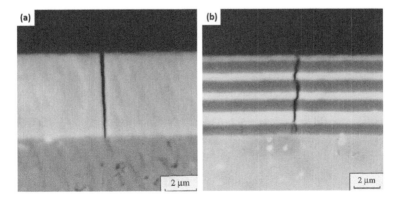

FIGURE 6.7 (a) Crack path in a monolayered coating NbN (b) Crack path in a multilayered coating TiN/NbN **Source:** For details see "Wiklund, U., Hedenqvist, P., & Hogmark, S. (1997). Multilayer cracking resistance in bending. Surface and Coatings Technology, 97(1–3), 773–778. https://doi.org/10.1016/S0257-8972(97)00290-9" Reprinted with permission from Elsevier

as shown in Figure 6.7 (b) (Wiklund et al., 1997). Furthermore, it has been reported that a noticeable reduction in residual stresses can be seen in multilayered TiN/Ti coating when compared to the monolayered TiN coating possibly due to the deformation of the soft Ti sublayer (Xie et al., 2022). Similar residual stress reduction was reported when TiAlN and TiAlCN were deposited as multilayers (Tillmann et al., 2021). As the coating properties depend upon the coating architecture, machining performance of the coatings will also be significantly affected by the architecture. Thus, in this section the effect of thin-film architecture on the machining performance of cutting tools will be discussed.

6.2.1 Coating characterization

In Section 6.1, it was seen that not only the coating thickness but also the monolayer and multilayer structure had a significant impact on the coating properties. Table 6.2 shows the properties of various coatings having monolayered, multilayered, and nanolayered coatings. It can be seen that the hardness of the multilayered coating in this case is higher when compared to its single-layered counterpart. Furthermore, the increase of layers leads to an increase in the hardness with the nanolayered structure showing the maximum hardness. This happens basically because of the superlattice effect when the multilayers have a nanoscale period thickness (Ducros et al., 2003; Hovsepian et al., 2006). However, the critical load and surface roughness values clearly indicate no specific relationship between the coating architecture,

TABLE 6.2
Coating properties for different monolayered, multilayered, and nanolayered coatings (Ducros et al., 2003)

Coating type	Metallic interlayer composition	Number of layers/coating thickness (μm)	Surface roughness, R_a (nm)	Hardness (HV0.05)	Critical load, L_C (N)-failure mode
TiN	Ti	1 layer/4±0.1	200	2500±100	25-cohesive 65-adhesive
CrN	Cr	1 layer/4±0.1	115	2300±100	30-cohesive 80-adhesive
Multilayer CrN/TiN	Cr	26 layer/3.7±0.1	140	2750±200	25-cohesive 75-adhesive
Multilayer CrN/TiN	Cr	52 layer/3.5±0.1	135	2900±200	25-cohesive 75-adhesive
Nanolayer CrN/TiN	Cr	3.8±0.01	160	3700±300	20-cohesive 65-adhesive
Multilayer TiN/AlTiN	Ti	26 layer/4±0.01	200	3000±200	20-cohesive 55-adhesive
Nanolayer TiN/AlTiN	Ti	4±0.01	210	3900±300	18-cohesive 50-adhesive

Effect of coatings on machining parameters 71

coating/substrate adhesion strength, and surface roughness. Also, as discussed earlier (see Figure 6.7), crack deflection is also an important phenomenon taking place in multilayered coatings which deflects the crack from its travel from the surface to the substrate.

6.2.2 Machining performance

As seen in Section 6.2.1, the coating architecture has a significant impact on some coating properties and thus a study of the machining performance of cutting tools while machining difficult-to-cut materials would be worthwhile. An interesting study as depicted in Figure 6.8 has been reported that shows a comparison between super hard monolayered TiAlSiN coating and a multilayered TiAlSiN coating having alternate layers of TiAlSiN with high and low hardness (Li et al., 2018). The study shows that hardness plays a dominant role in defining the performance of cutting tools. However, an optimized multilayered coating can outperform a harder monolayered coating (see Figure 6.9). In this case it was reported that the coatings suffer alternate shock loads while machining Inconel 718 superalloy which makes toughness a serious concern for the harder coating. Whereas the multilayered coating with alternate softer and harder phases has sufficient imparted toughness due to the presence of a softer TiAlSiN phase which helps it to dissipate the alternating shock waves. Also, there are similar studies that show superior machining performance of cutting tools with multilayer coatings (Kumar & Patel, 2018; Prengel et al., 2001).

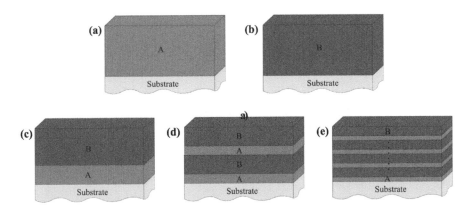

FIGURE 6.8 Schematic representation of monolayered and multilayered TiAlSiN coatings (a) TiAlSiN monolayer-soft (Coating A) (b) TiAlSiN monolayer-hard (Coating B) (c) bilayer (Multilayer-2 #M2) (d) 4 layer (Multilayer-4 #M4) and (e) 8 layer (Multilayer-8 #M8) **Source:** For details see "Li, G., Sun, J., Xu, Y., Xu, Y., Gu, J., Wang, L., Huang, K., Liu, K., & Li, L. (2018). Microstructure, mechanical properties, and cutting performance of TiAlSiN multilayer coatings prepared by HiPIMS. Surface and Coatings Technology, 353(June), 274–281. https://doi.org/10.1016/j.surfcoat.2018.06.017" Reprinted with permission from Elsevier

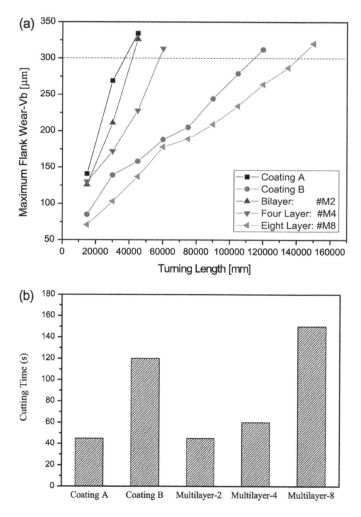

FIGURE 6.9 (a) Tool flank wear Vb curve versus cutting length for different TiAlSiN coated cutting tools; (b) tools life of flank wear above 300 μm of different coated tools (for notations see Figure 6.8) **Source:** For details see "Li, G., Sun, J., Xu, Y., Xu, Y., Gu, J., Wang, L., Huang, K., Liu, K., & Li, L. (2018). Microstructure, mechanical properties, and cutting performance of TiAlSiN multilayer coatings prepared by HiPIMS. Surface and Coatings Technology, 353(June), 274–281. https://doi.org/10.1016/j.surfcoat.2018.06.017" Reprinted with permission from Elsevier

6.2.3 Wear Mechanism

The wear mechanism of coated cutting tools helps in identifying the performance of coatings while machining different materials and under different cutting conditions. Figures 6.10 and 6.11 show the tool wear of the rake surface of the monolayered AlCrN and multilayered AlTiN-coated Al2O3/TiCN mixed ceramic cutting tools developed during the dry turning of hardened AISI 52100 steel at 63 HRC

Effect of coatings on machining parameters

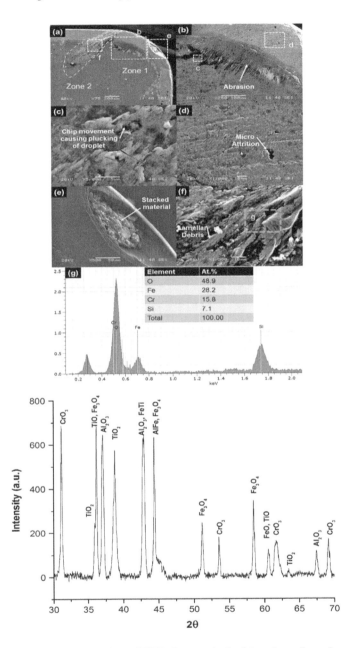

FIGURE 6.10 SEM micrographs and XRD phase analysis of the rake surface of monolayered AlCrN-coated cutting tool after machining **Source:** For details see "Kumar, C. S., & Patel, S. K. (2018). Performance analysis and comparative assessment of nano-composite TiAlSiN/TiSiN/TiAlN coating in hard turning of AISI 52100 steel. Surface and Coatings Technology, 335(September 2017), 265–279. https://doi.org/10.1016/j.surfcoat.2017.12.048" Reprinted with permission from Elsevier

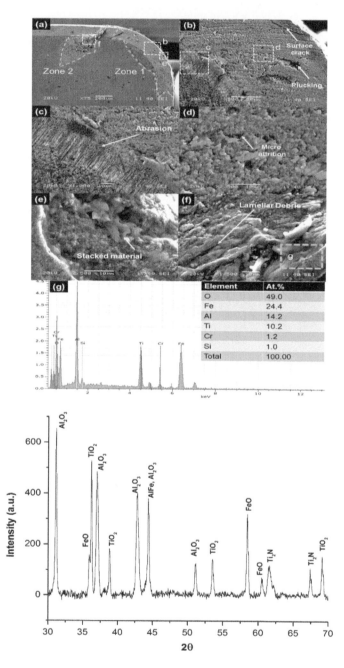

FIGURE 6.11 SEM micrographs and XRD phase analysis of the rake surface of multilayered AlTiN-coated cutting tool after machining **Source:** For details see "Kumar, C. S., & Patel, S. K. (2018). Performance analysis and comparative assessment of nano-composite TiAlSiN/TiSiN/TiAlN coating in hard turning of AISI 52100 steel. Surface and Coatings Technology, 335(September 2017), 265–279. https://doi.org/10.1016/j.surfcoat.2017.12.048" Reprinted with permission from Elsevier

hardness (Kumar & Patel, 2018). As can be seen, two wear zones have been identified on the rake face of the cutting tools. Zone 1, highlighted using a red dashed line, represents the wear region near the cutting edge whereas Zone 2, highlighted by a yellow dashed line, represents the wear region at the chip-tool interface. For the AlCrN-coated tool, tool wear occurring in Zone 1 is characterized by abrasion, edge chipping, and micro-attritions. The removal of coating from the surface of the cutting tools leads to the formation of microgrooves on the surface of the substrate. This microgroove formation occurs because thin-film takes away small pieces of the substrate during the delamination process (Neves et al., 2013). Furthermore, in Zone 2, the wear is characterized by chip sticking and seizure leading to the formation of a built-up layer (BUL). Also, the XRD phase analysis shows the formation of TiO_2 and CrO_3 indicating oxidation of the cutting tool and coating materials.

On the other hand, the wear in Zone 1 for the multilayered AlTiN coating is characterized by abrasion and attrition. There was no edge chipping or mechanical failure of the cutting edge as was the case with the AlCrN-coated tool. However, coating delamination leading to micro-attrition can be seen. Also, the adverse thermo-mechanical conditions prevailing during hard machining lead to the generation of surface cracks. Also, it can be seen that Zone 2, which is characterized by seizure and chip sticking, reduced in size when compared to the AlCrN-coated tool, indicating superior anti-adhesive properties for the AlTiN coating. From the above discussion it is evident that the multilayered coating with almost similar mechanical properties can outperform a monolayered coating.

6.2.4 CONCLUSIONS

The above discussions regarding the effect of coating structure/architecture on the machining performance of cutting tools show that multilayered coatings can develop several beneficial properties like crack deflection and superlattice effects leading to higher mechanical stability during machining tests. However, the coating structure has to be optimized for lower residual stresses, higher hardness, and better toughness. Also, it was observed that harder coatings exhibit higher wear resistance but during machining difficult-to-cut materials like Inconel 718 it is necessary to impart certain toughness to the coatings so that the shock loads can be absorbed.

6.3 PERFORMANCE OF COATED TOOLS WITH LUBRICATION

Coatings can be very effective in reducing friction and providing superior wear resistance for cutting tools during machining (Barshilia et al., 2010; Endrino et al., 2006; Kumar Sahoo & Sahoo, 2013). However, there can be certain situations where the machining conditions are very adverse leading to severe coating delamination (Sateesh et al., 2020). In such cases, application of lubrication during machining can be helpful in reducing the adverse effects of the machining conditions and taking advantage of the functions offered by the coatings at the same time. The cutting fluids offer functions such as chip disposal, improved machining accuracy and surface finish, and an extension of tool life. However, the disposal cost of used cutting fluids is high and due to the health and environmental disadvantages associated with cutting fluids, their use is often regulated.

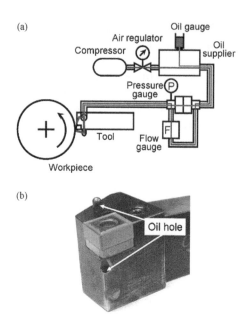

FIGURE 6.12 Oil supply system for MQL setup **Source:** For details see "Kamata, Y., & Obikawa, T. (2007). High speed MQL finish-turning of Inconel 718 with different coated tools. Journal of Materials Processing Technology, 192–193, 281–286. https://doi.org/10.1016/j.jmatprotec.2007.04.052" Reprinted with permission from Elsevier

In this regard, minimum quantity lubrication (MQL) offers certain benefits like reduced lubrication and disposal costs (Klocke & Eisenblätter, 1998). In order to understand the effect of lubrication on the machining performance of cutting tools during the machining of difficult-to-cut materials, high-speed finish turning of Inconel 718 using different coated tools under dry and MQL conditions will be discussed (Kamata & Obikawa, 2007). MQL is a lubrication technique in machining where lubricant is sprayed on the tool tip with the help of compressed air. The literature shows that the MQL technique results in a similar or superior improvement in tool life and surface integrity when compared to traditional flood cooling techniques (Chinchanikar & Choudhury, 2014; Sayuti et al., 2014; Shokrani et al., 2017; Yazid et al., 2011). Figure 6.12 demonstrates a schematic representation of the MQL oil supply system.

6.3.1 Tool wear under dry, wet, and MQL conditions

In the present example, three different coatings are considered, namely TiCN/Al2O3/TiN (Coating A), TiN/AlN (Coating B), and TiAlN (Coating C). The coatings were deposited on the cemented carbide cutting tools. All the coated tools were subjected to continuous turning tests under dry, wet, and MQL conditions. Figure 6.13 shows the corner wear variation with cutting length for all the coated tools under dry, wet, and MQL conditions respectively. It has been observed that the wear rates for Coating A reduce under wet and MQL conditions. However, for Coatings B and C, the wear rate increases in a wet environment, indicating adhesion between tool and workpiece

Effect of coatings on machining parameters

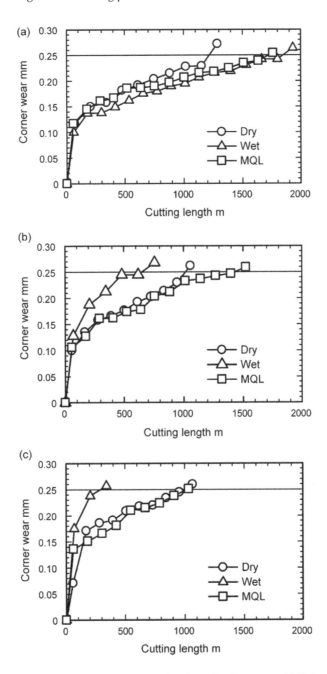

FIGURE 6.13 Tool wear for different coated tools under dry, wet and MQL conditions (a) Coating A (b) Coating B and (c) Coating C **Source:** For details see "Kamata, Y., & Obikawa, T. (2007). High speed MQL finish-turning of Inconel 718 with different coated tools. Journal of Materials Processing Technology, 192–193, 281–286. https://doi.org/10.1016/j.jmatprotec.2007.04.052" Reprinted with permission from Elsevier

at low cutting temperatures. However, MQL and a dry cutting environment account for almost similar wear rates for Coatings B and C. Thus, from the results it is evident that the wet cutting environment can have a negative impact if the tool and workpiece have strong adhesive properties at lower cutting temperatures. This problem can be addressed by keeping the quantity of lubricant in the cutting zone at a minimum.

6.3.2 Tool life and surface integrity

The tool life and surface integrity (Baowan et al., 2017; Chinchanikar & Choudhury, 2015; Hood et al., 2013) are the most important factors that define the performance of deposited thin-films. Figure 6.14 shows the variation of tool life and surface roughness (R_a) for different coated cutting tools under dry, wet, and MQL conditions

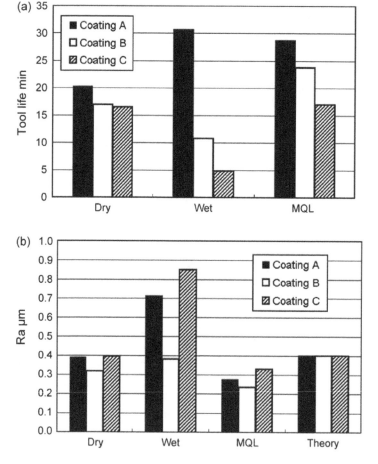

FIGURE 6.14 (a) Tool life and (b) surface roughness for different coated cutting tools under dry, wet and MQL cutting conditions **Source:** For details see "Kamata, Y., & Obikawa, T. (2007). High speed MQL finish-turning of Inconel 718 with different coated tools. Journal of Materials Processing Technology, 192–193, 281–286. https://doi.org/10.1016/j.jmatprotec .2007.04.052" Reprinted with permission from Elsevier

respectively. The cutting tool coated with Coating A accounted for a higher tool life under all conditions. The wear rate discussed earlier has a direct impact on the tool life of the cutting tools. As the wear rate increased for tools coated with Coatings B and C under a wet environment, the tool life reduced drastically when compared to dry working conditions. However, an improvement in tool life can be seen for all the coated cutting tools for the MQL condition. Further, the MQL conditions for cutting tools accounted for the lowest surface roughness values. Thus, the MQL conditions can help not only in improving the tool life but also in maintaining the surface integrity of the machined part at desired levels.

6.3.3 Conclusions

The present section dealt with the effect of lubrication on the performance of coated cutting tools during the machining of difficult-to-cut materials – in this case Inconel 718. From the presented results it is evident that the use of lubrication during machining can be helpful in improving the tool life and surface quality of the machined surface. However, the quantity of lubrication has to be optimized for the best outcomes. Thus, the MQL cutting environment can be a more appropriate lubrication technique that can be implemented with coated cutting tools.

REFERENCES

Baowan, P., Saikaew, C., & Wisitsoraat, A. (2017). Influence of helix angle on tool performances of TiAlN– and DLC-coated carbide end mills for dry side milling of stainless steel. *International Journal of Advanced Manufacturing Technology*, 90(9–12), 3085–3097. https://doi.org/10.1007/s00170-016-9601-5

Barshilia, H. C., Ghosh, M., Shashidhara, Ramakrishna, R., & Rajam, K. S. (2010). Deposition and characterization of TiAlSiN nanocomposite coatings prepared by reactive pulsed direct current unbalanced magnetron sputtering. *Applied Surface Science*, 256(21), 6420–6426. https://doi.org/10.1016/j.apsusc.2010.04.028

Chinchanikar, S., & Choudhury, S. K. (2014). Hard turning using HiPIMS-coated carbide tools: Wear behavior under dry and minimum quantity lubrication (MQL). *Measurement: Journal of the International Measurement Confederation*, 55, 536–548. https://doi.org/10.1016/j.measurement.2014.06.002

Chinchanikar, S., & Choudhury, S. K. (2015). Machining of hardened steel – Experimental investigations, performance modeling and cooling techniques: A review. *International Journal of Machine Tools and Manufacture*, 89, 95–109. https://doi.org/10.1016/j.ijmachtools.2014.11.002

Ducros, C., Benevent, V., & Sanchette, F. (2003). Deposition, characterization and machining performance of multilayer PVD coatings on cemented carbide cutting tools. *Surface and Coatings Technology*, 163–164, 681–688. https://doi.org/10.1016/S0257-8972(02)00656-4

Endrino, J. L., Fox-Rabinovich, G. S., & Gey, C. (2006). Hard AlTiN, AlCrN PVD coatings for machining of austenitic stainless steel. *Surface and Coatings Technology*, 200(24), 6840–6845. https://doi.org/10.1016/j.surfcoat.2005.10.030

Hood, R., Aspinwall, D. K., Sage, C., & Voice, W. (2013). High speed ball nose end milling of γ-TiAl alloys. *Intermetallics*, 32, 284–291. https://doi.org/10.1016/j.intermet.2012.09.011

Hovsepian, P. E., Luo, Q., Robinson, G., Pittman, M., Howarth, M., & Doerwald, D. (2006). TiAlN / VN superlattice structured PVD coatings : A new alternative in machining of aluminium alloys for aerospace and automotive components. *Surface and Coatings Technology*, 201, 265–272. https://doi.org/10.1016/j.surfcoat.2005.11.106

Jiang, F., Yan, L., & Rong, Y. (2013). Orthogonal cutting of hardened AISI D2 steel with TiAlN-coated inserts: Simulations and experiments. *International Journal of Advanced Manufacturing Technology, 64*(9–12), 1555–1563. https://doi.org/10.1007/s00170-012-4122-3

Kamata, Y., & Obikawa, T. (2007). High speed MQL finish-turning of Inconel 718 with different coated tools. *Journal of Materials Processing Technology, 192–193*, 281–286. https://doi.org/10.1016/j.jmatprotec.2007.04.052

Klocke, F., & Eisenblätter, G. (1998). Dry cutting – State of research. *VDI Berichte, 46*(1399), 159–188.

Kong, X., Yang, L., Zhang, H., Zhou, K., & Wang, Y. (2015). Cutting performance and coated tool wear mechanisms in laser-assisted milling K24 nickel-based superalloy. *International Journal of Advanced Manufacturing Technology, 77*(9–12), 2151–2163. https://doi.org/10.1007/s00170-014-6606-9

Kumar, C. S., & Patel, S. K. (2017). Surface & coatings technology experimental and numerical investigations on the effect of varying AlTiN coating thickness on hard machining performance of Al2O3-TiCN mixed ceramic inserts. *SCT, 309*, 266–281. https://doi.org/10.1016/j.surfcoat.2016.11.080

Kumar, C. S., & Patel, S. K. (2018). Performance analysis and comparative assessment of nano-composite TiAlSiN/TiSiN/TiAlN coating in hard turning of AISI 52100 steel. *Surface and Coatings Technology, 335*(September 2017), 265–279. https://doi.org/10.1016/j.surfcoat.2017.12.048

Kumar Sahoo, A., & Sahoo, B. (2013). Performance studies of multilayer hard surface coatings (TiN/TiCN/Al2O3/TiN) of indexable carbide inserts in hard machining: Part-II (RSM, grey relational and techno economical approach). *Measurement, 46*(8), 2868–2884. https://doi.org/10.1016/j.measurement.2012.09.023

Li, G., Sun, J., Xu, Y., Xu, Y., Gu, J., Wang, L., Huang, K., Liu, K., & Li, L. (2018). Microstructure, mechanical properties, and cutting performance of TiAlSiN multilayer coatings prepared by HiPIMS. *Surface and Coatings Technology, 353*(June), 274–281. https://doi.org/10.1016/j.surfcoat.2018.06.017

Neves, D., Eduardo, A., Sérgio, M., & Lima, F. (2013). Applied surface science microstructural analyses and wear behavior of the cemented carbide tools after laser surface treatment and PVD coating. *Applied Surface Science, 282*, 680–688. https://doi.org/10.1016/j.apsusc.2013.06.033

Posti, E., & Nieminen, I. (1989). Influence of coating thickness on the life of TiN-coated high speed steel cutting tools. *Wear, 129*(2), 273–283. https://doi.org/10.1016/0043-1648(89)90264-0

Prengel, H. G., Jindal, P. C., Wendt, K. H., Santhanam, A. T., Hegde, P. L., & Penich, R. M. (2001). A new class of high performance PVD coatings for carbide cutting tools. *Surface and Coatings Technology, 139*(1), 25–34. https://doi.org/10.1016/S0257-8972(00)01080-X

Qin, F., Chou, Y. K., Nolen, D., & Thompson, R. G. (2009). Coating thickness effects on diamond coated cutting tools. *Surface and Coatings Technology, 204*(6–7), 1056–1060. https://doi.org/10.1016/j.surfcoat.2009.06.011

Sargade, V. G., Gangopadhyay, S., Paul, S., & Chattopadhyay, A. K. (2011). Effect of coating thickness on the characteristics and dry machining performance of TiN film deposited on cemented carbide inserts using CFUBMS. *Materials and Manufacturing Processes, 26*(8), 1028–1033. https://doi.org/10.1080/10426914.2010.526978

Sateesh, C., Majumder, H., Khan, A., & Kumar, S. (2020). Applicability of DLC and WC/C low friction coatings on Al2O3/TiCN mixed ceramic cutting tools for dry machining of hardened 52100 steel. *Ceramics International, November 2019*, 0–1. https://doi.org/10.1016/j.ceramint.2020.01.225

Sateesh Kumar, C., & Kumar Patel, S. (2017). Hard machining performance of PVD AlCrN coated Al2O3/TiCN ceramic inserts as a function of thin film thickness. *Ceramics International*, *43*(16), 13314–13329. https://doi.org/10.1016/j.ceramint.2017.07.030

Sateesh Kumar, C., Majumder, H., Khan, A., & Patel, S. K. (2020). Applicability of DLC and WC/C low friction coatings on Al2O3/TiCN mixed ceramic cutting tools for dry machining of hardened 52100 steel. *Ceramics International*, *46*(8), 11889–11897. https://doi.org/10.1016/j.ceramint.2020.01.225

Sayuti, M., Sarhan, A. A. D., & Salem, F. (2014). Novel uses of SiO2 nano-lubrication system in hard turning process of hardened steel AISI4140 for less tool wear, surface roughness and oil consumption. *Journal of Cleaner Production*, *67*, 265–276. https://doi.org/10.1016/j.jclepro.2013.12.052

Shokrani, A., Dhokia, V., & Newman, S. T. (2017). Hybrid cooling and lubricating technology for CNC milling of inconel 718 nickel alloy. *Procedia Manufacturing*, *11*(June), 625–632. https://doi.org/10.1016/j.promfg.2017.07.160

Thakur, A., Gangopadhyay, S., & Mohanty, A. (2015). Investigation on some machinability aspects of inconel 825 during dry turning. *Materials and Manufacturing Processes*, *30*(8), 1026–1034. https://doi.org/10.1080/10426914.2014.984216

Tillmann, W., Grisales, D., Stangier, D., Thomann, C. A., Debus, J., Nienhaus, A., & Apel, D. (2021). Residual stresses and tribomechanical behaviour of TiAlN and TiAlCN monolayer and multilayer coatings by DCMS and HiPIMS. *Surface and Coatings Technology*, *406*(November 2020). https://doi.org/10.1016/j.surfcoat.2020.126664

Toparli, M., Sen, F., Culha, O., & Celik, E. (2007). Thermal stress analysis of HVOF sprayed WC-Co/NiAl multilayer coatings on stainless steel substrate using finite element methods. *Journal of Materials Processing Technology*, *190*(1–3), 26–32. https://doi.org/10.1016/j.jmatprotec.2007.03.115

Vereschaka, A. A., Grigoriev, S. N., Volosova, M. A., Batako, A., Vereschaka, A. S., Sitnikov, N. N., & Seleznev, A. E. (2017). Nano-scale multi-layered coatings for improved efficiency of ceramic cutting tools. *International Journal of Advanced Manufacturing Technology*, *90*(1–4), 27–43. https://doi.org/10.1007/s00170-016-9353-2

Wiklund, U., Hedenqvist, P., & Hogmark, S. (1997). Multilayer cracking resistance in bending. *Surface and Coatings Technology*, *97*(1–3), 773–778. https://doi.org/10.1016/S0257-8972(97)00290-9

Xie, W., Zhao, Y., Liao, B., Wang, S., & Zhang, S. (2022). Comparative tribological behavior of TiN monolayer and Ti/TiN multilayers on AZ31 magnesium alloys. *Surface and Coatings Technology*, *441*(March), 128590. https://doi.org/10.1016/j.surfcoat.2022.128590

Yazid, M. Z. A., Ibrahim, G. A., Said, A. Y. M., CheHaron, C. H., & Ghani, J. A. (2011). Surface integrity of Inconel 718 when finish turning with PVD coated carbide tool under MQL. *Procedia Engineering*, *19*, 396–401. https://doi.org/10.1016/j.proeng.2011.11.131

7 Significance of finite element analysis in analyzing performance of coated cutting tools

Finite element analysis is used extensively for studying the machining performance of cutting tools. It is a useful tool when studying chip formation mechanism for numerically predicting adiabatic shear band formation (ASB) (Uçak et al., 2020), chip serration (Jomaa et al., 2017), stresses developed in the chips during machining (Umbrello et al., 2008), strain along the shear plane (Yen et al., 2004), and temperature distribution in both tool and workpiece materials (Kumar & Patel, 2017a). Further, numerical prediction of the performance of coated cutting tools during machining of difficult-to-cut materials has proved very effective in understanding the performance of different coatings (Kumar & Patel, 2017b; Materials et al., 2018; Outeiro et al., 2008; Sateesh Kumar & Kumar Patel, 2017).

Figure 7.1 shows a simplified simulation setup that represents boundary conditions, coating layers, indicative finite element mesh, and machining parameters used for performing simulations. The setup shows a simulation setup that is used for finite element analysis of the turning process. However, an actual setup with a cylindrical workpiece can also be used in place of a simplified linear model but it will increase the complexity and solving time of the problem due to more curved parts involved in the numerical calculations. Also, a linear model prevents meshing errors and reduces computational time. Further, 2D models can be used to precisely study the chip formation process that involve the study of the formation of adiabatic shear band (ASB), stresses in the chip, serrated teeth formation, and chip thickness (Fan & Li, 2009; Jomaa et al., 2017; Kumar et al., 2020). Also, the application of finite element analysis for predicting the machining performance of coated cutting tools could be very helpful to understand various tool wear mechanisms like adhesion, abrasion, and diffusion which are very important during the machining of difficult-to-cut materials. In this regard, the application of finite element analysis techniques for predicting chip formation mechanisms, machining forces, stress distribution, and temperature distribution during the machining of difficult-to-cut materials using cutting tools with thin-film depositions will be discussed in the subsequent sections.

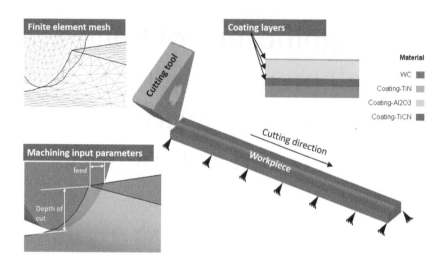

FIGURE 7.1 Simplified linear setup for 3D turning simulations

7.1 STUDY OF MACHINING FORCES AND STRESS DISTRIBUTION

Prediction of forces using finite element techniques is quite significant in metal cutting operations as its modeling helps in performing tool life estimation, temperature distribution analysis, and chip formation studies (Huang & Liang, 2003). However, accurate prediction of machining forces and stress distribution using finite element techniques significantly depends on the proper definition of workpiece and cutting tool material properties such as flow stress of the workpiece material, failure criterion for the workpiece material, thermal conductivity, emissivity, heat capacity, and so on (Coelho et al., 2007; Kumar et al., 2020; Sateesh Kumar & Kumar Patel, 2017). Also, it has been revealed that finite element analysis and modeling techniques can be very helpful in predicting cutting forces while machining with coated cutting tools (Yadav et al., 2015). Furthermore, stress distribution in the tool and workpiece materials can also be predicted using finite element numerical techniques. Stress distribution analysis helps in understanding sensitive areas on the tool and workpiece material that directly influence the tool life and surface quality of the machined surface (Kumar & Patel, 2017a; Özel, 2006; Sateesh Kumar & Kumar Patel, 2017). Based on the above discussion, the present section discusses the numerical studies that elaborate on the prediction of machining forces and stress distribution in the tool and workpiece material using finite element techniques.

7.1.1 Machining forces

It is well known that the measurement or prediction of machining forces can be very helpful in precisely understanding the shearing process of metal cutting, tool wear mechanism, and coefficient of friction at the chip tool interface (Wyen & Wegener, 2010). Higher machining forces can be detrimental to tool wear and the surface quality of the machined surface. The higher machining forces lead to high cutting zone temperatures eventually elevating the tool wear rates and poor surface finish (Sateesh Kumar & Kumar Patel, 2017). With all these benefits of measuring

Significance of finite element analysis

machining forces during a metal cutting operation, sometimes it becomes necessary to validate a numerical finite element model through machining forces. As discussed earlier, forces during machining act as a source of determining the performance of cutting tools. Further, the study of forces becomes even more significant as the coatings are expected to act as a thermal barrier between the cutting tool substrate and the workpiece, provide a lubrication effect by reducing friction, and improve wear resistance for prolonged cutting tool durability.

In this regard, Figure 7.2 shows the variation of cutting force, thrust force, and feed force with the thickness of AlCrN coating (Sateesh Kumar & Kumar Patel, 2017). The experiments and simulations were carried out with the help of Al_2O_3/TiCN-based mixed ceramic cutting tools. AISI 52100 alloy steel hardened to 63 HRC hardness was used as the workpiece for the tests.

No lubrication was used during the machining process. It can be seen that the deposition of the coating on the mixed ceramic substrate results in the reduction of machining forces. Further, the most significant output that we can draw from these results is that the finite element modeling and analysis were able to predict the machining forces corresponding to coated tools accurately. Although there was never an exact match between the measured and predicted machining forces, the numerically predicted machining forces followed the trend suggested by the experimentally measured forces using a piezoelectric dynamometer. However, it was reported that the finite element model was unable to predict the drastic increase in machining forces while machining

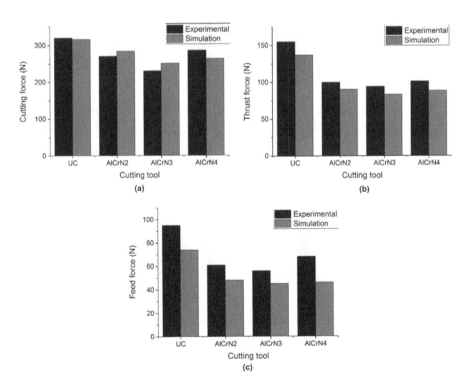

FIGURE 7.2 Variation of machining forces with the coating thickness of AlCrN coating on Al_2O_3/TiCN mixed ceramic cutting tool

with the cutting tool having 4 μm coating thickness AlTiN coating. This extreme rise in machining forces was attributed to the lower coating/substrate adhesion strength and higher edge radius. It has been reported by various researchers that the effect of edge radius (Jagadesh & Samuel, 2014; Komanduri et al., 1998; Nasr et al., 2007; Özel & Ulutan, 2012) can be studied using finite element analysis. On the other hand, analysis of the effect of coating/substrate adhesion strength on the machining performance of coated cutting tools has not been reported yet in the literature. Furthermore, when considering cutting tools as rigid in the finite element model, it becomes impossible to define interface properties.

Figure 7.3 also shows the comparison of machining forces that were experimentally measured and numerically predicted respectively during the machining of

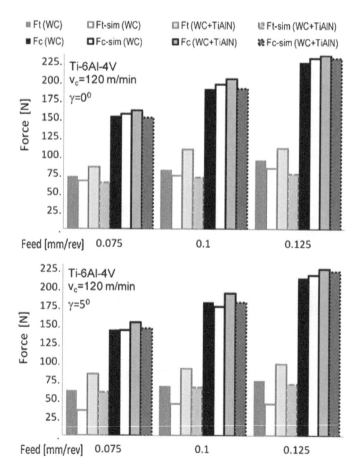

FIGURE 7.3 Variation of machining forces with feed rate for different uncoated and coated cutting tools at (a) 0 ° and (b) 5 ° rake angle **Source**: For details see "Özel, T., Sima, M., Srivastava, A. K., & Kaftanoglu, B. (2010). Investigations on the effects of multi-layered coated inserts in machining Ti-6Al-4V alloy with experiments and finite element simulations. CIRP Annals – Manufacturing Technology, 59(1), 77–82. https://doi.org/10.1016/j.cirp.2010.03.055". Reprinted with permission from Elsevier

Ti-6Al-4V alloy (Özel et al., 2010). These experiments were performed with the help of uncoated tungsten carbide (WC/Co) and coated tungsten carbide cutting tools under a dry-cutting environment. The coatings that were considered are TiAlN, cBN (cubic-boron nitride), and a duplex coating of cBN/TiAlN. The simulation results were validated with the measured orthogonal machining forces (cutting force (F_C) and thrust force (F_t)). The results clearly indicate a good agreement between the experimental and simulation results. Further, it can be concluded from the above discussion that the finite element modeling and analysis techniques are capable of studying the performance of coated cutting tools having different thin-film depositions.

7.1.2 STRESS DISTRIBUTION

As discussed earlier, stress distribution analysis in both the tool and the workpiece can be very effective in identifying critical areas with high-stress concentration. This analysis can help in finding not only the reasons for high or low tool wear but also help in reasoning more specific wear mechanisms such as chipping, mechanical breakage, abrasion, and material adhesion (Arrazola & Özel, 2010; Chinchanikar & Choudhury, 2015; Kumar & Patel, 2017a). Furthermore, residual stress analysis can also be performed for different machining operations carried out using coated cutting tools (Özel & Ulutan, 2012). In this regard, Figures 7.4 and 7.5 show the machining-induced residual stresses in Ti-6Al-4V titanium alloy and IN100 superalloy respectively. The machining-induced residual stresses were measured using X-ray diffraction techniques. Both circumferential and radial residual stresses were studied. The simulations were able to predict the circumferential residual stresses more accurately when compared to that in the radial direction. In addition, the simulations were very effective in studying the effect of coatings on residual stresses. The tensile residual stresses reduced slightly for TiAlN coated tools which was revealed by both experimental and predicted results. Also, the simulations were able to predict the variation of circumferential stresses with good accuracy for both uncoated and coated cutting tools.

The stresses in cutting tools can be very helpful in predicting tool wear and wear mechanisms. The stress distribution in the cutting tools can be significantly affected by tool geometry and working conditions (Kumar & Patel, 2017a; Kurt et al., 2015; Kurt & Şeker, 2005; Lazoglu et al., 2006). In this regard, Figure 7.6 shows the stress distribution in the cutting tools while machining Inconel 625 superalloy. It was reported that the effective stress increased with the increase in depth of cut. Furthermore, the wear rate was higher at higher stress concentration points. Thus, from the above discussion, it can be concluded that the finite element techniques adopted for predicting effective stress distribution on the cutting tools help in predicting critical wear points that may have high wear rates during actual machining experiments (Lotfi et al., 2016). Also, these stress distribution results help classify more critical wear mechanisms such as edge chipping, mechanical breakage, and adhesion by identifying high-stress concentration points. These results are often correlated with cutting tool wear taking place during actual experiments. Similarly, the finite element predictive techniques are also capable of predicting stress and strain very precisely in the chips formed during the metal cutting operation which will be discussed in the subsequent sections.

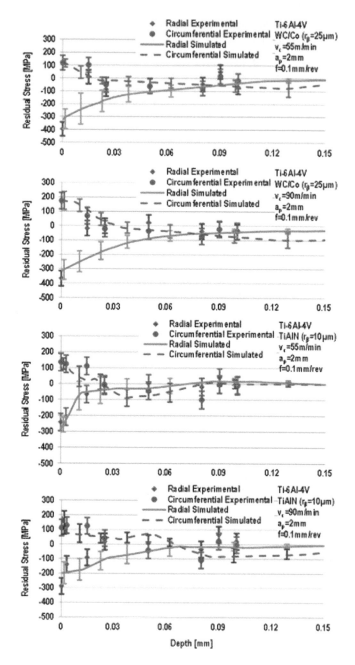

FIGURE 7.4 Machining-induced residual stress comparison in Ti-6Al-4V titanium alloy while machining with uncoated and TiAlN coated cutting tools **Source**: For details see "Özel, T., & Ulutan, D. (2012). Prediction of machining induced residual stresses in turning of titanium and nickel based alloys with experiments and finite element simulations. CIRP Annals – Manufacturing Technology, 61(1), 547–550. https://doi.org/10.1016/j.cirp.2012.03.100". Reprinted with permission from Elsevier

Significance of finite element analysis

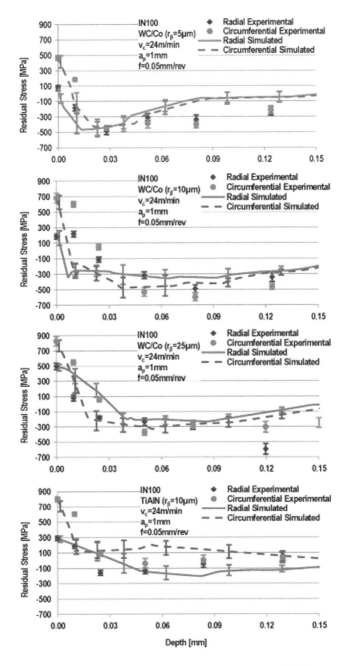

FIGURE 7.5 Machining-induced residual stress comparison in IN100 superalloy while machining with uncoated and coated cutting tools **Source**: For details see "Özel, T., & Ulutan, D. (2012). Prediction of machining induced residual stresses in turning of titanium and nickel based alloys with experiments and finite element simulations. CIRP Annals – Manufacturing Technology, 61(1), 547–550. https://doi.org/10.1016/j.cirp.2012.03.100". Reprinted with permission from Elsevier

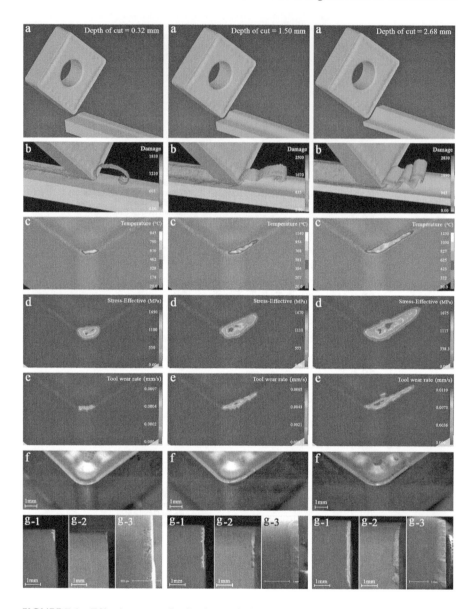

FIGURE 7.6 Effective stress distribution variation with depth of cut in coated cutting tools
Source: For details see "Lotfi, M., Jahanbakhsh, M., & Akhavan Farid, A. (2016). Wear estimation of ceramic and coated carbide tools in turning of Inconel 625: 3D FE analysis. Tribology International, 99, 107–116. https://doi.org/10.1016/j.triboint.2016.03.008". Reprinted with permission from Elsevier

7.2 STUDY OF TEMPERATURE DISTRIBUTION

Temperature distribution study in the workpiece and cutting tools during machining can be very helpful in understanding the thermal aspects of machining especially thermal deterioration of cutting tools. It is well known that at high temperatures the cutting tools lose their favorable cutting properties like high hardness, oxidation resistance, and wear resistance. This happens because high temperatures can affect the chemical bonding within the materials during machining which may lead to deterioration of properties (Kumar & Patel, 2017a; Yang et al., 2011). Furthermore, the temperature distribution investigations using finite element techniques become more significant as the exact cutting zone is never exposed properly due to the complex processes taking place during machining. One such complex phenomenon is the movement of the chips over the tool face which keeps the exact cutting zone hidden and makes it very difficult to explore it using the measuring devices. Thus, the measured temperatures using contact and non-contact devices don't give a clear picture of the effect of the temperature on the tool wear, and surface quality of the machined surface. Thus, finite element techniques have been extensively used to study the temperature distribution during metal cutting operations (Grzesik et al., 2005; Hargrove & Ding, 2007; Hung et al., 2018; Kumar & Patel, 2017a; Yang et al., 2011; Zhang et al., 2019).

As discussed earlier, the machining of difficult-to-cut materials like hardened steels and superalloys may lead to high cutting temperatures which makes it essential to investigate the temperature distribution to get a clear understanding of the effect of thermal aspects on machining output. In this regard, Figure 7.7 shows the temperature distribution in the uncoated and coated Al_2O_3/TiCN mixed ceramic cutting tools having different thin-film thicknesses of AlTiN multi-layered coating. The machining simulations were performed taking into consideration a dry-cutting environment using the Deform 3D software package. AISI 52100 steel having 63 HRC hardness was taken as workpiece material for all the simulations. The numerical results revealed that the deposition of AlTiN coating resulted in a reduction in tool temperatures. Furthermore, it has been reported in the literature that the coatings act as a thermal barrier and prevent heat to flow to the substrate (Kumar et al., 2020) which has also been observed in the temperature distribution results. In addition, another important temperature is the interface temperature where the chip slides over the cutting tool. This is the area where the maximum temperature is generated due to the combined effect of friction and adhesion.

Figure 7.8 shows the variation of tool and interface temperatures while machining hardened AISI 52100 steel. A comparison of the measured and predicted temperatures has also been shown. From the results, it is evident that the measured and the simulated results had a difference of 10 to 20%. Also, the simulated temperatures cannot be considered with an error as the temperature measurement techniques are not capable of measuring the tool and interface temperatures very accurately due to the complexities discussed earlier.

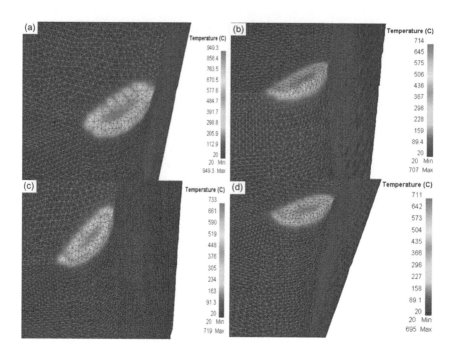

FIGURE 7.7 Finite element predicted temperature distribution in (a) uncoated, (b) 2 μm AlTiN coated, (c) 3 μm AlTiN coated, and (d) 4 μm AlTiN coated Al2O3/TiCN mixed ceramic cutting tools **Source**: For details see "Kumar, C. S., & Patel, S. K. Experimental and numerical investigations on the effect of varying AlTiN coating thickness on hard machining performance of Al2O3-TiCN mixed ceramic inserts. Surface & Coatings Technology, (2017b) 309, 266–281. https://doi.org/10.1016/j.surfcoat.2016.11.080". Reprinted with permission from Elsevier

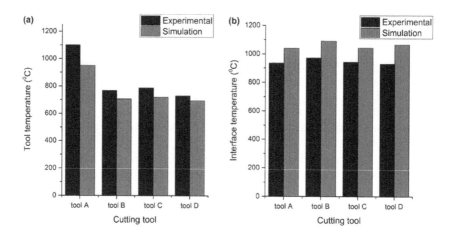

FIGURE 7.8 Finite element predicted and experimentally measured (a) tool temperature and (b) interface temperature for uncoated and AlTiN coated mixed ceramic cutting tools with varying thin-film thickness from 2 to 4 μm

7.2.1 CONCLUSIONS

From the above discussions, it can be concluded that the finite element techniques are highly effective in investigating the temperature distribution in cutting tools and workpiece materials. Also, these investigations can assist in understanding the tool wear mechanism and surface quality of the machined surface. Further, the simulations were also able to predict the effect of coatings and coating thickness on the temperature distribution in both tool and workpiece material. Thus, finite element techniques are very helpful in understanding the thermal barrier effect of coating depositions on the cutting tools (Kumar et al., 2020; Uçak et al., 2020).

7.3 CHIP MORPHOLOGY ANALYSIS

The chip formation mechanism provides information about the machining performance of cutting tools and suggests the applicability of cutting tools during a particular metal cutting operation. As discussed earlier, the compressive force exerted by the cutting tool on the workpiece generates plastic strain along a localized region called the shear plane. The progress of the machining processes leads to the intensification of this strain component leading to the formation of an adiabatic shear band and chip segmentation (Jomaa et al., 2017; Kumar et al., 2020). The high-temperature generation at the chip tool contact zone leads to the formation of serrated teeth. So, the machining of materials like Inconel, hardened steels, and other super alloys may lead to the formation of serrated teeth. Also, studying various chip geometrical parameters like chip thickness, chip width, and serrated teeth height are very helpful in analyzing the tool wear and surface quality of the machined surface. However, to accurately comprehend the variation in these geometrical parameters due to changes in machining conditions, tool material, and workpiece properties is only possible using finite element techniques that can provide data related to stress and strain developed during the chip formation and the temperature distribution within the chip during machining.

Thus, the use of finite element techniques is very helpful in understanding the chip formation mechanism while machining difficult-to-cut materials using coated cutting tools. In this regard, Figure 7.9 shows a comparison between experimental and simulated chips formed while machining Ti-6Al-4V titanium alloy using coated carbide cutting tools. The machining results in the formation of chips with serrated teeth. Also, as shown in the figure, the simulated results show a good agreement with the chips formed while machining. Further, a close comparison of the minimum and maximum serrated teeth height proves this experimental and simulated results agreement (Özel et al., 2010).

The finite element techniques could also be very helpful in studying the chip flow direction that can be quantified using chip bend angles. Figure 7.10 shows the chip bend angles for the chips formed while machining AISI 52100 hardened steel using uncoated and AlTiN-coated mixed ceramic cutting tools. In this case, higher chip bend angles indicate higher friction. If the friction is higher the chips

FIGURE 7.9 Finite element predicted and actual serrated chip formation comparison
Source: For details see "Özel, T., Sima, M., Srivastava, A. K., & Kaftanoglu, B. (2010). Investigations on the effects of multi-layered coated inserts in machining Ti-6Al-4V alloy with experiments and finite element simulations. CIRP Annals – Manufacturing Technology, 59(1), 77–82. https://doi.org/10.1016/j.cirp.2010.03.055". Reprinted with permission from Elsevier

tend to stick on the tool surface and thus, they will not move away from the tool face resulting in higher chip bend angles. Alternatively, lower friction allows free movement of the chips away from the tool face leading to lower chip bends in this case.

7.3.1 Conclusions

From these discussions, it is evident that the finite element techniques are not only able to predict chip geometrical parameters like chip thickness, chip width, the maximum and minimum height of serrated teeth, and serration frequency, but also are quite capable of predicting chip flow direction by quantifying chip bend angles. Also, these numerical methods provide proper information on stresses, strains, and temperature distribution developed within the chips that help in understanding the chip formation mechanisms such as the development of adiabatic shear bands in serrated teeth.

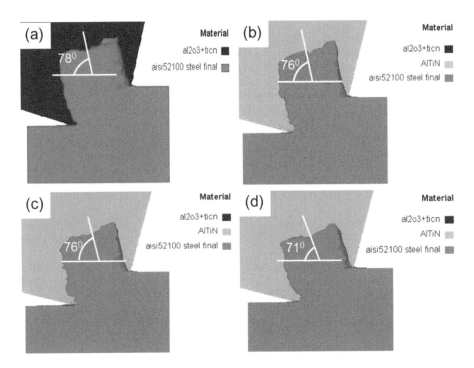

FIGURE 7.10 Numerically predicted chip bend angles while machining hardened AISI 52100 steel using uncoated and AlTiN coated mixed ceramic cutting tools **Source:** For details see "Kumar, C. S., & Patel, S. K. (2017b). Experimental and numerical investigations on the effect of varying AlTiN coating thickness on hard machining performance of Al 2 O 3 -TiCN mixed ceramic inserts. Surface & Coatings Technology, 309, 266–281. https://doi.org /10.1016/j.surfcoat.2016.11.080". Reprinted with permission from Elsevier

REFERENCES

Arrazola, P. J., & Özel, T. (2010). Investigations on the effects of friction modeling in finite element simulation of machining. *International Journal of Mechanical Sciences*, 52(1), 31–42. https://doi.org/10.1016/j.ijmecsci.2009.10.001

Chinchanikar, S., & Choudhury, S. K. (2015). Machining of hardened steel – Experimental investigations, performance modeling and cooling techniques: A review. *International Journal of Machine Tools and Manufacture*, 89, 95–109. https://doi.org/10.1016/j.ijmachtools.2014.11.002

Coelho, R. T., Ng, E. G., & Elbestawi, M. A. (2007). Tool wear when turning hardened AISI 4340 with coated PCBN tools using finishing cutting conditions. *International Journal of Machine Tools and Manufacture*, 47(2), 263–272. https://doi.org/10.1016/j.ijmachtools.2006.03.020

Fan, W. F., & Li, J. H. (2009). An investigation on the damage of AISI-1045 and AISI-1025 steels in fine-blanking with negative clearance. *Materials Science and Engineering A*, 499(1–2), 248–251. https://doi.org/10.1016/j.msea.2007.11.108

Grzesik, W., Bartoszuk, M., & Nieslony, P. (2005). Finite element modelling of temperature distribution in the cutting zone in turning processes with differently coated tools. *Journal of Materials Processing Technology*, *165*, 1204–1211. https://doi.org/10.1016/j.jmatprotec.2005.02.136

Hargrove, S. K., & Ding, D. (2007). Determining cutting parameters in wire EDM based on workpiece surface temperature distribution. *International Journal of Advanced Manufacturing Technology*, *34*(3–4), 295–299. https://doi.org/10.1007/s00170-006-0609-0

Huang, Y., & Liang, S. Y. (2003). Cutting forces modeling considering the effect of tool thermal property??? Application to CBN hard turning. *International Journal of Machine Tools and Manufacture*, *43*(3), 307–315. https://doi.org/10.1016/S0890-6955(02)00185-2

Hung, T.-P., Shi, H.-E., & Kuang, J.-H. (2018). Temperature modeling of AISI 1045 steel during surface hardening processes. *Materials*, *11*(10), 1815. https://doi.org/10.3390/ma11101815

Jagadesh, T., & Samuel, G. L. (2014). Finite element modeling for prediction of cutting forces during micro turning of titanium alloy. *Aimtdr*, 6–11. https://doi.org/10.1080/10910344.2015.1085318

Jomaa, W., Mechri, O., Lévesque, J., Songmene, V., Bocher, P., & Gakwaya, A. (2017). Finite element simulation and analysis of serrated chip formation during high–speed machining of AA7075–T651 alloy. *Journal of Manufacturing Processes*, *26*(October), 446–458. https://doi.org/10.1016/j.jmapro.2017.02.015

Komanduri, R., Chandrasekaran, N., & Raff, L. M. (1998). Effect of tool geometry in nanometric cutting: A molecular dynamics simulation approach. *Wear*, *219*(1), 84–97. https://doi.org/10.1016/S0043-1648(98)00229-4

Kumar, C. S., & Patel, S. K. (2017a). Effect of WEDM surface texturing on Al2O3/TiCN composite ceramic tools in dry cutting of hardened steel. *Ceramics International*, *44*(2), 2510–2523. https://doi.org/10.1016/j.ceramint.2017.10.236

Kumar, C. S., & Patel, S. K. (2017b). Experimental and numerical investigations on the effect of varying AlTiN coating thickness on hard machining performance of Al 2 O 3 -TiCN mixed ceramic inserts. *Surface & Coatings Technology*, *309*, 266–281. https://doi.org/10.1016/j.surfcoat.2016.11.080

Kumar, C. S., Zeman, P., & Polcar, T. (2020). A 2D finite element approach for predicting the machining performance of nanolayered TiAlCrN coating on WC-Co cutting tool during dry turning of AISI 1045 steel. *Ceramics International*, *46*(16), 25073–25088. https://doi.org/10.1016/j.ceramint.2020.06.294

Kurt, A., & Şeker, U. (2005). The effect of chamfer angle of polycrystalline cubic boron nitride cutting tool on the cutting forces and the tool stresses in finishing hard turning of AISI 52100 steel. *Materials and Design*, *26*(4), 351–356. https://doi.org/10.1016/j.matdes.2004.06.022

Kurt, A., Yal??in, B., & Yilmaz, N. (2015). The cutting tool stresses in finish turning of hardened steel with mixed ceramic tool. *International Journal of Advanced Manufacturing Technology*, 315–325. https://doi.org/10.1007/s00170-015-6927-3

Lazoglu, I., Buyukhatipoglu, K., Kratz, H., & Klocke, F. (2006). Forces and temperatures in hard turning. *Machining Science and Technology*, *10*(2), 157–179. https://doi.org/10.1080/10910340600713554

Lotfi, M., Jahanbakhsh, M., & Akhavan Farid, A. (2016). Wear estimation of ceramic and coated carbide tools in turning of Inconel 625: 3D FE analysis. *Tribology International*, *99*, 107–116. https://doi.org/10.1016/j.triboint.2016.03.008

Materials, H., Kumar, C. S., & Patel, S. K. (2018). International journal of refractory metals application of surface modification techniques during hard turning: Present work and future prospects. *International Journal of Refractory Metals & Hard Materials*, *76*(March), 112–127. https://doi.org/10.1016/j.ijrmhm.2018.06.003

Nasr, M. N. A., Ng, E. G., & Elbestawi, M. A. (2007). Modelling the effects of tool-edge radius on residual stresses when orthogonal cutting AISI 316L. *International Journal of Machine Tools and Manufacture*, *47*(2), 401–411. https://doi.org/10.1016/j.ijmachtools.2006.03.004

Outeiro, J. C., Pina, J. C., M'Saoubi, R., Pusavec, F., & Jawahir, I. S. (2008). Analysis of residual stresses induced by dry turning of difficult-to-machine materials. *CIRP Annals – Manufacturing Technology*, *57*(1), 77–80. https://doi.org/10.1016/j.cirp.2008.03.076

Özel, T. (2006). The influence of friction models on finite element simulations of machining. *International Journal of Machine Tools and Manufacture*, *46*(5), 518–530. https://doi.org/10.1016/j.ijmachtools.2005.07.001

Özel, T., & Ulutan, D. (2012). Prediction of machining induced residual stresses in turning of titanium and nickel based alloys with experiments and finite element simulations. *CIRP Annals – Manufacturing Technology*, *61*(1), 547–550. https://doi.org/10.1016/j.cirp.2012.03.100

Özel, T., Sima, M., Srivastava, A. K., & Kaftanoglu, B. (2010). Investigations on the effects of multi-layered coated inserts in machining Ti-6Al-4V alloy with experiments and finite element simulations. *CIRP Annals – Manufacturing Technology*, *59*(1), 77–82. https://doi.org/10.1016/j.cirp.2010.03.055

Sateesh Kumar, C., & Kumar Patel, S. (2017). Hard machining performance of PVD AlCrN coated Al2O3/TiCN ceramic inserts as a function of thin film thickness. *Ceramics International*, *43*(16), 13314–13329. https://doi.org/10.1016/j.ceramint.2017.07.030

Uçak, N., Aslantas, K., & Çiçek, A. (2020). The effects of Al2O3 coating on serrated chip geometry and adiabatic shear banding in orthogonal cutting of AISI 316L stainless steel. *Journal of Materials Research and Technology*, *9*(5), 10758–10767. https://doi.org/10.1016/j.jmrt.2020.07.087

Umbrello, D., Ambrogio, G., Filice, L., & Shivpuri, R. (2008). A hybrid finite element method-artificial neural network approach for predicting residual stresses and the optimal cutting conditions during hard turning of AISI 52100 bearing steel. *Materials and Design*, *29*(4), 873–883. https://doi.org/10.1016/j.matdes.2007.03.004

Wyen, C. F., & Wegener, K. (2010). Influence of cutting edge radius on cutting forces in machining titanium. *CIRP Annals – Manufacturing Technology*, *59*(1), 93–96. https://doi.org/10.1016/j.cirp.2010.03.056

Yadav, R. K., Abhishek, K., & Mahapatra, S. S. (2015). A simulation approach for estimating flank wear and material removal rate in turning of Inconel 718. *Simulation Modelling Practice and Theory*, *52*, 1–14. https://doi.org/10.1016/j.simpat.2014.12.004

Yang, K., Liang, Y. C., Zheng, K. N., Bai, Q. S., & Chen, W. Q. (2011). Tool edge radius effect on cutting temperature in micro-end-milling process. *International Journal of Advanced Manufacturing Technology*, *52*(9–12), 905–912. https://doi.org/10.1007/s00170-010-2795-z

Yen, Y.-C., Jain, A., Chigurupati, P., Wu, W.-T., & Altan, T. (2004). Computer simulation of orthogonal cutting using a tool with multiple coatings. *Machining Science and Technology*, *8*(2), 305–326. https://doi.org/10.1081/MST-200029230

Zhang, S., Zhang, W., Wang, P., Liu, Y., Ma, F., Yang, D., & Sha, Z. (2019). Simulation of material removal process in EDM with composite tools. *Advances in Materials Science and Engineering*, *2019*, 1–11. https://doi.org/10.1155/2019/1321780

8 Conclusions

1. The chemical, mechanical, and thermal properties of different materials make them difficult to cut under different situations. The demand for advanced materials in challenging sectors like defense, aerospace, battery manufacturing, medical equipment, and manufacturing dealing with high-temperature applications with superior wear resistance, heat resistance, and chemical and thermal stability under adverse conditions led to the development of materials like hardened steels, titanium alloys, superalloys, new ceramics, and composites. The high surface hardness, hot hardness, low thermal conductivity, and chemical affinity are some of the properties that generate difficulty in removing material by metal cutting.
2. The material for cutting tools should be selected judiciously based on the applicability, precision, and surface quality requirements for the metal cutting process. As a general rule, the material of the cutting tool should be harder than the material to be cut. Further, while machining difficult-to-cut materials, the prerequisites for the cutting tools are high hot hardness, chemical, and thermal stability at high temperatures, and high toughness. Figure 1.8 shows different types of cutting tool materials as a function of hardness and toughness. It is evident from the illustration that for cutting tools toughness and hardness have an inverse relationship. The cutting tools with higher hardness like polycrystalline diamond (PCD) tools exhibit lower toughness. Furthermore, the cutting tools made of high-speed steel are extremely tough but at the same time exhibit very low levels of hardness when compared to other harder cutting tools made of polycrystalline cubic boron nitrides (PCBN) and ceramics. Thus, it becomes highly necessary to understand the use of different cutting tool materials for machining difficult-to-cut materials.
3. Taking into consideration the machining of difficult-to-cut materials, metal cutting operations with different machining difficulties and challenges. The metal cutting operation of materials like hardened steel and superalloys having low thermal conductivity generates high machining forces and cutting temperatures. This cutting environment leads to thermal deterioration and excessive abrasion of the cutting tools causing a substantial increase in tool wear rates. Further, at higher cutting temperatures the surface quality of the machined surface can also deteriorate due to chip sticking, and built-up-edge formation.
4. Various methods are being employed for tacking machining challenges and difficulties. The techniques focus on the reduction of friction and temperature during the metal cutting operation. The reduction in friction eventually has a cumulative effect on the metal cutting outcomes. The

reduced friction results in a reduction in machining forces which in turn reduces cutting temperatures. Thus, a significant reduction of friction during the machining operation not only results in an exponential increase in the cutting tool's durability but also improves the surface quality of the machined surface substantially. To achieve these cooling and lubrication functions, cutting fluids, solid lubricants, and thin-film depositions on the cutting tools are effectively employed. Further, cutting fluids possess significant disadvantages such as the generation of hazardous vapors due to the interaction of fluids and the heated machined surface, tool, and chips. These vapors give rise to different environmental and health hazards for workers. On the other hand, the use of solid lubricants can solve the hazardous effects of conventional cutting fluids up to some extent, but still, the problem is not completely eliminated. In addition, solid lubricants sticking to the workpiece and the machine tool calls for additional investment in cleaning. These disadvantages can be eliminated by the use of thin-film depositions on the cutting tools.

5. For the deposition of thin-films different technologies can be implemented. In general chemical vapor deposition (CVD) and physical vapor deposition (PVD) techniques are used for the deposition of coatings on cutting tools. The PVD techniques can be further classified into different categories like sputtering and arc evaporation. However, sputtering technology has seen various advancements in the form of balanced and unbalanced magnetron sputtering, high impulse magnetron sputtering (HiPIMS), and other hybrid techniques like HiPIMS+arc and conventional PVD+HiPIMS.

6. The thin-films can be classified into different categories based on their properties, architecture/structure, and chemical composition. The coatings based on their material can be classified into oxides, nitrides, carbides, etc. Further, based on the coating architecture the coatings can be classified as monolayer, multi-layer, nanocomposite, nanolayer, etc. Also, a general classification of the coatings can be made based on their properties like hard and soft coatings.

7. For improving the machinability and the cutting tool durability during the machining of difficult-to-cut materials like titanium alloys, nickel-based superalloys, ceramics, composites, and hardened steels, thin-film depositions on the cutting tools have proved to be very effective. The coatings offer various functions like generation of lubricious phases (e.g., V_2O_5, WO_3, etc.) during the machining operation resulting in the reduction of friction and machining forces, high hardness resulting in superior wear resistance due to the formation of the nanocomposite structure (e.g., amorphous Si_3N_4 and TiAlN forms TiAlSiN nanocomposite structure with very high hardness), and multi-layered structure causing crack deflection and reduction of residual stresses thereby improving the service life of the coatings. These functions offered by coatings are extremely significant when machining difficult-to-cut materials as they help in reducing tool wear rates and improve the surface quality of the machined surface.

Conclusions

8. It can be seen that the coating thickness significantly influences the thin-film characteristics that are reflected clearly in the performance of the coatings. Further, monolayer and multi-layered architecture also significantly impact the way the coatings behave during machining when the coating thickness is changed.
9. Investigations on the effect of coating structure/architecture on the machining performance of cutting tools show that the multilayered coatings can develop several beneficial properties like crack deflection and superlattice effects leading to higher mechanical stability during the machining tests. However, the coating structure has to be optimized for lower residual stresses, higher hardness, and better toughness. Also, it was observed that harder coatings exhibit higher wear resistance but during machining difficult-to-cut materials like Inconel 718, it is necessary to impart certain toughness to the coatings so that the shock loads can be absorbed.
10. From the discussed results, it is evident that the use of lubrication during machining can help improve the tool life and surface quality of the machined surface. However, the quantity of lubrication has to be optimized for the best outcomes. Thus, the MQL cutting environment can be a more appropriate lubrication technique that can be implemented with coated cutting tools.
11. From the discussion of the results, it is evident that the finite element techniques are not only able to predict chip geometrical parameters like chip thickness, chip width, the maximum and minimum height of serrated teeth, and serration frequency, but also are quite capable of predicting chip flow direction by quantifying chip bend angles. Also, these numerical methods provide proper information on stresses, strains, and temperature distribution developed within the chips that help in understanding the chip formation mechanisms such as the development of adiabatic shear bands in serrated teeth.

Index

Abrasion, 7, 8, 17, 21, 75, 83, 87, 99
Adhesion, 19, 21, 32, 39, 45, 55, 59, 63–65, 68, 69, 71, 76, 83, 86, 87, 91
Aerospace, 7, 8, 18, 41, 56, 99
Application, 43, 49
Architecture, 23, 43, 45, 63, 69, 70, 75, 100, 101

Barrier, 5, 22, 49, 55, 63, 85, 91
Bilayer, 69, 71
Built-up-edge, 2, 3, 18, 58

Carbides, 11, 31, 37, 43, 100
Carbon, 10, 37, 45, 46
Carbo-nitrides, 43, 44
Cathode, 27, 28
Ceramics, 9, 11, 65
Challenges, 17–19, 21, 23, 99
Chamber, 27, 29, 31, 33–35, 38, 39
Chemical reaction, 7, 33, 35
Chemical stability, 12, 18, 57, 63
Chemical vapor deposition (CVD), 33–38, 50–52, 100
Chip formation, 1, 2, 4, 17, 18, 41, 83, 84, 93, 94, 101
Chip morphology, 6, 66
CNC, 5
Coated tool, 75
Coating architecture, 49, 63, 70, 71, 100
Coating thickness, 63–70, 85, 86, 92, 95, 100
Composite, 9, 31, 47, 73, 74
Composition, 27, 31, 33, 43–45, 49, 70, 100
Conventional, 10, 22, 23, 29–33, 100
Coolant, 1, 6, 17, 22, 43
Crack, 23, 38, 46, 69, 71, 75, 100, 101
Cubic boron nitride (CBN), 12
Cutting fluids, 22, 23, 75, 100
Cutting force, 1, 67, 85, 87
Cutting operation, 1, 5, 6, 17, 19, 21–23, 53, 85, 87, 93, 99
Cutting speed, 1, 19, 20, 41, 49–51, 53–57
Cutting zone, 91

Deflection, 23, 71, 75, 100, 101
Deposition process, 30, 39, 47
Deposition techniques, 29, 63
Design, 45, 46
Diamond-like carbon (DLC), 23, 53, 54, 58, 64
Difficult-to-cut materials, 3, 7, 8, 10, 11, 17, 18, 21, 23, 49, 55, 58, 63, 71, 75, 76, 79, 83, 91, 93, 99–101

Disadvantages, 17, 22, 23, 31–33, 59, 75, 100
Dry machining, 1
Durability, 1, 7, 17, 21–23, 49, 85, 100

Environment, 6, 17, 23, 41, 49, 76, 78, 79, 87, 91, 99, 101

Failure, 18, 32, 41, 70, 75, 84
Fatigue, 8, 18
Finite element, 83, 92, 94
Friction, 17, 21, 22, 27, 41, 43–45, 49, 50, 53–55, 57, 58, 63, 66, 75, 84, 85, 91, 93, 94, 99, 100

Gases, 29, 31, 33–35, 37, 38
Geometry, 4, 5, 87
Graphite, 22, 37, 49, 53

Hard coating, 17, 43, 55
High impulse magnetron sputtering (HiPIMS), 30–33, 39, 43, 71, 72, 100
Hybrid deposition processes, 39

Industry, 1, 4, 8, 27, 41
Interlayer, 70
Ionization, 27–30, 32, 38

Kinetic, 29, 32

Limitations, 58
Load, 70
Low-friction coatings, 49, 53–55, 57, 58
Lubrication, 17, 22, 23, 41, 63, 75, 76, 79, 85, 100, 101

Machining, 1–8, 10–12, 17–23, 27, 41–45, 49–53, 55–59, 63, 65, 66, 69–76, 79, 83–89, 91–95, 99–101
Machining forces, 1, 6–8, 17, 18, 21, 22, 49, 50, 53, 63, 83–87, 99, 100
Machining parameters, 1, 3, 17, 43, 63, 83
Manufacturing, 1, 4, 7, 11, 27, 43, 99
Matrix, 9, 47
Mechanical, 4, 6–8, 27, 33, 41, 43, 45, 50, 71, 72, 75, 87, 99, 101
Mechanism, 1, 2, 17–19, 27, 32, 50, 72, 83, 84, 93
Metal cutting, 4
Minimum quantity lubrication (MQL), 17, 22, 76–79, 101
Monolayer, 43, 45, 46, 69–71, 100, 101
Multi-layer, 100

103

Nanocomposite, 23, 43, 45, 47, 55, 100
Nanolayer, 100
Nitrides, 10, 12, 27, 31, 37, 43, 44, 99, 100

Operations, 4–6, 17, 22, 23, 58, 84, 87, 91, 99
Oxidation, 7, 9, 12, 27, 43, 45, 69, 75, 91
Oxides, 31, 37, 43, 100

Physical vapor deposition (PVD), 21, 27, 33, 39, 43, 56, 65, 100
Plasma, 27–31, 38
Polycrystalline, 10, 12, 99
Polycrystalline cubic boron nitride (PCBN), 10, 12, 99
Polymer, 9

Quality, 1, 4, 5, 10, 17–19, 21, 22, 27, 33, 41, 43, 49, 53, 79, 84, 91, 93, 99–101

Reactor, 35, 37, 38

Serrated teeth, 3, 4, 83, 93, 94, 101
Shear, 1, 3, 4, 46, 83, 93, 94, 101
Simulation, 83, 87
Soft coatings, 45, 55
Sputtering, 27–32, 100
Steels, 3, 7, 8, 11, 46, 59, 63, 91, 93, 99, 100
Stress distribution, 83, 84, 87, 90
Stresses, 8, 43, 64, 70, 75, 83, 87–89, 94, 100, 101

Structure, 20, 36, 43, 49, 55, 60, 63, 69, 70, 75, 100, 101
Superalloys, 3, 7–9, 17, 18, 41, 50, 55, 57–59, 71, 87, 89, 91, 99, 100
Surface morphology, 50, 51, 56, 63, 69

Temperature distribution, 83, 91–94, 101
Thermal stability, 7, 10, 18, 43, 58, 99
Thin-films, 4, 17, 19, 23, 51, 58, 63, 64, 66, 78, 100
Thornton diagram, 32
Tool wear, 6, 8, 9, 17, 19, 21, 41, 50, 53, 55, 57–59, 63, 68, 69, 72, 75, 76, 83, 84, 87, 91, 93, 99, 100
Toughness, 10, 11, 43–45, 55, 71, 75, 99, 101

Uncoated tool, 49, 57

Vacuum, 27, 38

Whiskers, 37
Workpiece, 1, 2, 5, 6, 19, 21–23, 49, 50, 58, 60, 63, 76, 78, 83–85, 87, 91, 93, 100

X-ray, 87
XRD, 73–75

Yield, 1, 27, 29

Zone, 3, 4, 17, 35, 36, 78, 84, 91, 93